THE NEW WORLD OF
POLICE
ACCOUNTABILITY

This book is dedicated to
HERMAN GOLDSTEIN
who understood these issues long before anyone else.

THE NEW WORLD OF
POLICE
ACCOUNTABILITY

SAMUEL WALKER
University of Nebraska at Omaha

SAGE Publications
Thousand Oaks ▪ London ▪ New Delhi

For information:

Sage Publications, Inc.
2455 Teller Road
Thousand Oaks, California 91320
E-mail: order@sagepub.com

Sage Publications Ltd.
1 Oliver's Yard
55 City Road
London EC1Y 1SP
United Kingdom

Sage Publications India Pvt. Ltd.
B-42, Panchsheel Enclave
Post Box 4109
New Delhi 110 017 India

Printed in the United States of America on acid-free paper.

Library of Congress Cataloging-in-Publication Data

Walker, Samuel, 1942-
The new world of police accountability / Samuel Walker.
    p. cm.
Includes bibliographical references and index.
ISBN 1-4129-0943-0 (cloth)—ISBN 1-4129-0944-9 (pbk.)
    1. Police misconduct—United States. 2. Police administration—United States.
I. Title.
HV8141.W3458 2005
363.2′2—dc22   2004019520

05   06   07   08   09   10   9   8   7   6   5   4   3   2   1

Acquiring Editor:	Jerry Westby
Project Editor:	Claudia A. Hoffman
Copy Editor:	Edward Meidenbauer
Typesetter:	C&M Digitals (P) Ltd.
Indexer:	Nara Wood

CONTENTS

PREFACE

—————•◦•—————

This book represents the culmination of about 15 years of intensive research and consulting on the issue of police accountability and 30 years of research and writing on policing in general. Looking back, it is astonishing how much has changed in this field just during these past 15 years. One of the most important programs described in this book—police auditors—was not even created until 1993. Early intervention systems, meanwhile, were in their infancy and had never been studied.

My own thinking about the issue of police accountability and how to achieve it has changed dramatically over this past decade and a half. One reflection of this is the change in some of the most important terminology I use. Back then, I referred to "civilian review boards"; today I refer to "citizen oversight." Only 7 years ago, I undertook a study of "early warning systems" in policing. Today I refer to "early intervention systems." These changes in terminology reflect important changes in my understanding of the institutions and programs at hand.

On another important issue—the one that closes this book, in fact—my thinking is profoundly mixed. On the one hand I am far more optimistic about the possibilities of meaningful police reform and for achieving genuine police accountability. I think we now have a reasonably good understanding of what needs to be done, of what strategies and tools are likely to be effective. The main goal of this book is to bring these insights to a broader audience. At the same time, however, I remain deeply skeptical about the possibilities of lasting change. The evidence for these thoughts is found in Chapter Seven. Police departments, like universities, private corporations, and all large bureaucracies, are extremely difficult to change. Although this book is cautiously optimistic, only time will tell whether that optimism is fully justified.

One thing is certain, however. This book sums up the enormous changes in the field of police accountability over the past 15 years. In light of that, it is safe to assume that the next 5 or 10 years will witness further dramatic changes. There will undoubtedly be new strategies and tools, together with new evidence on what works and what does not. The next 15 years will undoubtedly be as exciting as the past 15.

ACKNOWLEDGMENTS

T his book is dedicated to Herman Goldstein. Herman has greatly influenced my understanding of what the police do and how we might effectively control officer conduct to ensure compliance with the standards of a democratic society. Through a career that extends back almost half a century, he has shaped our thinking about the police more than any other single individual. As a member of the American Bar Foundation Survey in the mid-1950s, he helped to uncover the complexity of the police role and the pervasive exercise of discretion by police officers. He has wrestled with these issues ever since. There is a direct connection between his early work on police discretion and this book. Along the way, Herman developed the concept of problem-oriented policing, arguably the most important new idea related to how the police should address crime and disorder. I would like to thank Herman for his influence on my own career and for his immeasurable impact on the field of policing.

Many other people also contributed to the ideas expressed in this book. Merrick Bobb is simply the nation's leading expert on police accountability, and I have learned much from him. I have also learned much about policing through my many conversations with Teresa Guerrero-Daley, the San Jose Independent Police Auditor; Mike Gennaco, head of the Office of Independent Review for the Los Angeles Sheriff's Department; Richard Rosenthal, head of the Portland, Oregon, Independent Police Review office; Sam Pailca, director of the Office of Professional Accountability in Seattle; Ellen Ceisler-Green, head of the Office of Integrity and Accountability in Philadelphia; Pierce Murphy, the Boise Ombudsman; and Tristan Bonn, the Public Safety Auditor in Omaha. Richard Jerome, formerly of the U.S. Department of Justice, has also provided useful insights into policing.

This book is a greatly expanded version of the ideas I set forth in an article published in the *St. Louis University Law Public Law Review* (Vol. XXII, No. 1, 2003). That article, in turn, was based on the presentation I gave at a Symposium on New Approaches to Ensuring the Legitimacy of Police Conduct that the law school sponsored in April 2002. I would like to thank the law school and Professor Roger Goldman in particular for inviting me to speak at the conference and providing the opportunity to develop my ideas on this subject.

Several police departments have been particularly open and generous with their time. In particular I would like to thank officers in the Phoenix, AZ; Minneapolis, MN; and Austin, TX, police departments, as well as those in the Los Angeles Sheriff's Department who have helped me.

Geoff Alpert of the University of South Carolina has been a great friend and colleague on the subject of early intervention systems. Lorie Fridell, Research Director at the Police Executive Research Forum, has supported my work in various ways and has also been a great friend. Several of my former graduate students have also contributed to my work and the ideas in this book. Vic Bumphus undertook the first national survey of citizen oversight mechanisms for his master's thesis. Carol Archbold's dissertation investigated police risk management systems. Dawn Irlbeck has assisted with early intervention systems and other projects. Finally, Chuck Katz has been a great student and is now coauthor of another book with me.

ONE

INTRODUCTION TO THE NEW POLICE ACCOUNTABILITY

———●•●———

THE NEW POLICE ACCOUNTABILITY AT WORK: THREE EXAMPLES

Problems at the Century Station in the Los Angeles Sheriff's Department

The Century Station of the Los Angeles Sheriff's Department (LASD) was deeply troubled in the late 1990s. Officers assigned there averaged more than 12 shootings of citizens a year. Admittedly, the Century Station was a tough assignment, covering a high crime area in South Central Los Angeles County. Between 1991 and 2001, half of all LASD deputies killed by gunfire worked out of the Century Station. Merrick Bobb, Special Counsel to the LASD, described the Century Station as "a microcosm of American policing in inner city, crime-ridden, minority neighborhoods."[1] Closer inspection suggested there was nothing inevitable about the number of officer-involved shootings. The neighboring Los Angeles Police Department's Southeast precinct, with comparable social conditions, had only one third as many officer-involved shootings during the same years.[2]

An investigation by Special Counsel Bobb found that the issues in the Century Station were the result of management problems and not the fault of a few bad officers. Century Station deputies were among the youngest and least experienced in the entire LASD, and they were "supervised by an equally young and inexperienced crew of sergeants." Even more alarming, the department had allowed the ratio of sergeants to officers on the street to rise to an average of 1:10 or 1:12, and at times as high as 1:20–25, "in direct violation of the LASD's own official policy of 1:8."[3] The Century Station also suffered from an extremely high turnover rate among officers. Officers at all ranks wanted to transfer out as rapidly as possible and feared they would be stigmatized if they stayed there too long. Finally, the station had the additional burden of a very high percentage of probationary officers in training. They were assigned there in the—very misguided—belief that a high crime area, or what the LASD calls a "fast" assignment, would give them a lot of experience in a short period of time.[4]

Bobb's report led to a number of management reforms. The department assigned its best and "most-likely-to-be-promoted" lieutenants to the station, and they implemented tighter supervision of deputies on the street, with a special focus on potential shooting incidents. The results were dramatic. There was only one officer-involved shooting in the Century Station in 1999 and during one 17-month period there were no shootings at all. Significantly, the decline in shootings was not accompanied by a similar dip in the crime rate or number of arrests: "Rather," Bobb concluded, "it was directly related to increased supervision by the sergeants and lieutenants, along with a new effort to discourage foot pursuits, actions which were implicated in a large number of shootings at the station."[5]

Ending Racial Profiling by the New Jersey State Police

Sued by the Civil Rights Division of the U.S. Department of Justice for well-documented race discrimination in traffic stops, the New Jersey State Police entered into a consent decree mandating a set of management reforms. Most important was the requirement that troopers report each traffic stop to their dispatchers when they initiate the stop; that they activate a video and audio recording of each stop; and that they complete a detailed report of the encounter, including the reasons for the stop, the race or ethnicity of the driver and any passengers, and the final outcome (e.g., search, traffic citation, arrest). Additionally, the data on each traffic stop are entered into a computerized database that permits a detailed analysis of patterns and the identification of any unacceptable patterns of stops or searches.[6]

Systematic Performance Review in a Large Police Department

In a large urban police department in the western United States, the early intervention (EI) system database displayed some alarming patterns of officer activity. One officer had made only eight arrests during the time period under review (some other officers in the same unit had made more than 100) and had used force in five of those situations. A use of force rate of 63% is virtually unheard of in policing. Another officer, meanwhile, had a seemingly exemplary record of no citizen complaints and no use of force reports. Closer examination, however, revealed that he had made no arrests, no traffic stops, and no pedestrian stops during the period. In short, he had done no real police work. With a few clicks of the mouse, the Internal Affairs officers quickly found that he was working the maximum number of hours in off-duty employment (this information is also included in the department's early intervention system database). In short, all his energy was going into his second job and he was doing little if any police work for the city. A third officer had received a citizen complaint from a woman alleging an inappropriate sexual advance. A check of the officer's performance record revealed a suspiciously high number of traffic stops of female drivers. In short, he was a sexual predator using his law enforcement authority to harass women.[7]

These performance patterns were revealed by the department's EI system, an administrative tool that collects officer performance data in a computerized database in which they can be analyzed to identify patterns of questionable conduct. Officers identified by the system are then given appropriate counseling or retraining. Although EI systems were initially devised as a means of identifying officers who repeatedly use excessive force, the three examples here indicate that they can identify a wide range of officer performance problems.[8]

NEW DIRECTIONS IN POLICE ACCOUNTABILITY

The three cases described above are examples of the new police accountability: new strategies and tools for dealing with the ancient problems of police misconduct.[9] Merrick Bobb's investigation of LASD's Century Station found that the problem of shootings was not a matter of a few bad officers—the proverbial "rotten apples"—but poor management practices. Law professor Barbara Armacost argues that police reform needs to focus on "rotten barrels" rather than rotten apples.[10] Bobb's role as Special Counsel to the LASD represents an important new practice: permanent external oversight of a law enforcement agency by an expert in policing with a sizeable support staff. The new controls over traffic enforcement in the New Jersey State Police, meanwhile, involve the requirement that officers should be subject to detailed written rules of conduct and that they complete a report each time they use their authority, including use of force, a traffic stop, and deployment of the departmental canine unit. Finally, the early intervention system described in the third example illustrates how this new management tool can be used to identify a broad range of unacceptable performance, including doing no police work at all.

Police misconduct is nothing new in the United States. Use of excessive force (or what is popularly known as "police brutality"), unjustified shootings, race discrimination, and a general lack of accountability for officer conduct have been serious problems since the first police departments were created in the early nineteenth century. Several generations of reform efforts have attempted to curb these problems and establish professional standards in polic-ing, but until recently with only limited success.[11] As Chapter Two of this book argues, these reforms largely failed to address persistent on-the-street miscon-duct. In the past decade, however, new strategies and tools have emerged that promise to achieve new standards of police accountability. This book exam-ines the promise of—and the obstacles facing—the new police accountability.

The basic thrust of the new police accountability is a focus on *organiza-tional change*. This is a departure from past reform efforts that have focused too much on individual officers who may have used excessive force or made a racially biased arrest. The result has been a misplaced attention to symptoms (the rotten apples) rather than underlying organizational causes (the rotten barrels). As LASD Special Counsel Merrick Bobb explains,

> The basic premise on which we operate is that the risk of excessive and unnecessary force, lethal and non-lethal alike, can be meaningfully reduced through conscientious work on police management's part.[12]

Barbara Armacost, noting the depressing cycle of scandal, reform, and subsequent scandal in Los Angeles, argues that "reform efforts have focused too much on notorious incidents and misbehaving individuals," and not enough on police organizations that sustain a "police culture that facilitates and rewards violent conduct."[13]

Attention to organization and management problems is also the central thrust of the federal "pattern or practice" suits that have been brought against law enforcement agencies under Section 14141 of the 1994 Violent Crime Control Act.[14] The law specifically authorizes the U. S. Department of Justice to bring civil suits against law enforcement agencies, as opposed to criminal charges against individual officers. The consent decrees, memoranda of agreement, and letters settling the Department of Justice suits against the Pittsburgh Police Bureau (1997), the New Jersey State Police (1999), the Los Angeles Police Department (2001), and about a dozen other agencies require them to implement a set of organizational reforms. All of these settlements include the key elements of the new police accountability discussed in this book. They include a comprehensive use of force reporting system (Chapter Three), an open and accessible citizen complaint procedure (Chapter Four), and an early intervention system (Chapter Five). These "best practices" are described in the 2001 Department of Justice report *Principles for Promoting Police Integrity*, the first document to present them as a coherent package of police reforms.[15]

Implementation of each consent decree, meanwhile, is overseen by a court-appointed monitor who regularly reviews the police department's compliance with the terms of the decree and issues a public report on the findings. These monitors, however, exist only for the duration of the consent decree, and their authority is limited to the specific terms of the decree. Police auditors, a new form of citizen oversight described in Chapter Six, function on a permanent basis and are an additional element of the new police accountability.

THE GOAL OF THIS BOOK

The goal of this book is to describe and analyze the strategies and tools that constitute the new world of police accountability. Although this book argues that the strategies and tools of the new police accountability represent an important new development in policing and that they hold great promise for reducing police misconduct, it also gives due attention to the considerable obstacles facing the successful implementation of these new mechanisms. Indeed, Chapters Three through Six report substantial evidence of false starts and failures associated with the new accountability mechanisms. The book concludes in Chapter Seven with reflections on these problems and some thoughts on what is needed to fully achieve the potential of the new police accountability.

SOURCE MATERIAL

This book is built on four basic sets of source material: consent decrees and memoranda of understanding, reports of court-appointed monitors, reports of police auditors, and other reports of police problems.

Consent Decrees and Memoranda of Understanding

The various consent decrees and memoranda of understanding settling federal pattern or practice suits, along with the settlements of other cases, are basic source material on the new accountability mechanisms. By implication, they also document the specific problems in particular police departments. The fact that the Memorandum of Agreement in Cincinnati contains a provision requiring documentation of each incident in which police officers draw their weapons indicates that this practice occurred frequently and was perceived by the community to be a problem in that city.[16]

The Reports of Court-Appointed Monitors

Each consent decree includes the requirement of a court-appointed monitor who is responsible for regularly auditing progress in implementing the terms of the decree and then issuing a public report, usually on a quarterly basis. The reports that have been issued to date are an extremely valuable resource, particularly on the issue of organizational change in policing. Many people mistakenly assume that if a court orders a police department to, for example, implement a new use of force policy, the new policy will be promptly implemented. Whereas some of the departments subject to consent decrees have implemented mandated reforms in a timely fashion, others have not. Taken as a whole, the reports of all the court-appointed monitors provide a revealing picture of the difficulties of bringing about change in a law enforcement organization.[17]

The Reports of Police Auditors

As Chapter Six explains, police auditors have emerged as a new form of citizen oversight of the police. By mid-2004, there were twelve police auditors in the United States. Each auditor is required to issue public reports on a regular basis. These reports typically explore particular police problems and make recommendations for changes in policies and procedures. The particular value of the auditing approach to citizen oversight is the capacity of the auditor to conduct a follow-up investigation months or years later. The reports of the various police auditors provide a wealth of valuable information about problems within the various departments about policy changes and the process of organizational change.[18]

Other Reports on Police Problems

This book also draws on a variety of reports on police problems. Some of these are reports on special one-time investigations of a particular department or problem within a department. In addition, some police departments, as a part of their own accountability efforts, have created ongoing internal reviews of such issues as the use of force or racial profiling. Some of these efforts include substantial citizen participation and the publication of periodic reports. These reports are extremely valuable sources of information.[19]

Because they have emerged relatively recently, many of the documents listed above are unknown to many experts in the field of policing. Nonetheless, they are a good source of information about the key issues related to police accountability. Because we now live in the digital age, all of these documents are readily available on the web. The appropriate web addresses are cited throughout this book.

A DEFINITION OF POLICE ACCOUNTABILITY

It is appropriate at the outset to define what we mean by police accountability in this book. Police accountability has two basic dimensions.[20] On one level it refers to holding law enforcement *agencies* accountable for the basic services they deliver: crime control, order maintenance, and miscellaneous services to people and communities. At the same time, however, it also refers to holding *individual officers* accountable for how they treat individual citizens, particularly with regard to the use of force, equal treatment of all groups, and respect for the dignity of individuals. In certain important respects, of course, the agency-level and officer-level dimensions of accountability merge. Effective crime control and order maintenance depend on what individual officers do on the street. And individual officer misconduct ultimately depends on what police departments do to define and enforce standards of conduct.[21]

It is a basic principle of a democratic society that the police should be answerable to the public.[22] The political process is the basic means of ensuring that the police reflect the will of the people. Mayors, city council members, county commissioners, governors, state legislatures, presidents, and the Congress exercise control and oversight through budgets and appointments.[23] At the same time, the police are also accountable to the law and should conform to established standards of lawfulness in all of their operations (including not just law enforcement activities such as arrest and search and seizures, but personnel procedures involving equal employment opportunity, sexual harassment, and so on). The courts are the principal mechanism for this aspect of accountability.

In a democratic society the three branches of government are ultimately controlled by the citizenry, acting through the political process. The public also acts through other important nongovernmental institutions. Particularly important in recent American history has been the role of civil rights and civil liberties organizations. Local chapters of the NAACP and the ACLU, along with countless ad hoc community groups, have exerted an enormous impact on American policing. Finally, the news media have played a sporadic but nonetheless powerful influence on public understanding of police problems. The impact of the famous videotape of the 1991 beating of Rodney King by Los Angeles police officers is incalculable.

One of the greatest obstacles to police accountability in this country is that, while the general roles of the three branches of government are clear, the specific strategies and tools for ensuring police accountability involve complex administrative arrangements. The unhappy fact is that most elected officials do not understand the details of police administration and have been unable—and often unwilling—to provide guidance to law enforcement executives. Mayors in particular are rarely knowledgeable about policing and generally subject to a variety of countervailing political pressures, such as public demands to get tough on crime. Legislators are often less closely attuned to the details of police administration and prone to symbolic legislative gestures. Although the federal courts stepped into this vacuum in the 1960s and began to require detailed controls, the courts, as Chapter Two explains in more detail, are an imperfect tool for the ongoing administrative reform of police departments.

Nothing is easy in a democracy, however, and the influence of public opinion over the police has been a very mixed blessing. Public control of the

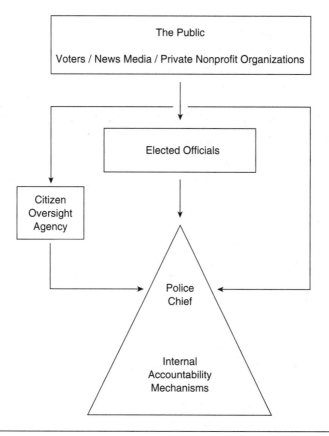

Figure 1.1   A graphic model of police accountability.

police and other government agencies has often involved the denial of rights to racial minorities and other powerless groups, as white majorities associate crime with people who are not like them.[24] Compounding the problem of accountability is the melancholy but inescapable fact that very often public demands for crime control conflict with principles of due process and equal protection of the law. The history of American criminal justice is replete with episodes in which public demands for law and order conflicted with the rights of powerless groups.[25] In a classic statement of the problem, Herbert Packer defined it in terms of a clash between crime control and due process perspectives on the criminal process.[26] The story of the Supreme Court's revolution in civil rights and due process in the late 1950s and 1960s involves the role of the most undemocratic political institution in our society in overriding the wishes of the majority to protect the rights of otherwise powerless minorities.

The political dilemma involving conflicting demands on the police translates into policy dilemmas over alternative mechanisms for enhancing police accountability. This has generally taken the form of tensions between internal mechanisms, such as professional management and supervision, and external mechanisms, such as direct political control, the courts, and external review agencies. This book argues that this traditional conflict is now sterile and irrelevant. The strategies and tools of the new police accountability involve a merger of internal and external mechanisms that have traditionally been seen as conflicting alternatives.

THE BEST OF TIMES, THE WORST OF TIMES: AMERICAN POLICING TODAY

Images of Police Misconduct

The images, broadcast nationwide on television on November 30, 2003, were depressingly familiar: police officers delivering repeated blows with billy clubs on Nathaniel Jones, an African American man, who died a few hours later. The fact that these were Cincinnati police officers immediately brought back memories of the riot that broke out in the city in April 2001 following the fifteenth fatal shooting of a young black man in 6 years by the Cincinnati police.[27]

The video of the 2003 Cincinnati beating inevitably recalled the sensational images of the March 3, 1991 beating of Rodney King by Los Angeles police officers. The subsequent acquittal of four LAPD officers on criminal charges in June 1992 sparked massive rioting and property destruction in Los Angeles and other cities.[28]

Not captured on any video are some equally alarming examples of police misconduct. In Miami, four officers (out of eleven indicted) were convicted in 2003 of beating and framing citizens. A civil suit arising from similar misconduct resulted in a consent decree imposing sweeping organizational reforms on the Oakland, California, Police Department. Finally, what began as a civil suit involving sexual abuse of a teenage girl by a Pennsylvania State Police officer led to the exposure in 2003 of a massive pattern of sexual misconduct by Pennsylvania state troopers, including several high-ranking command officers.

The police problems in Cincinnati, Los Angeles, Oakland, Miami, Pennsylvania, and other communities might well lead a reasonable person— that is, someone well-informed about civic events but with no special expertise in policing—to conclude that the American police have made little if any progress since the strife-torn decade of the 1960s.[29] Race discrimination, excessive force, unjustified shootings, and corruption might appear to be as prevalent and serious today as they were 40 years earlier. This might also suggest to our hypothetical reasonable person that the many police reform efforts of the past four decades have accomplished nothing. These efforts include celebrated Supreme Court decisions limiting abusive police practices;[30] the spread of community policing and problem-oriented policing;[31] a significant increase in the number of African American, Hispanic, and women police officers;[32] and dramatic improvements in police officer educational levels and training programs.[33] One might reasonably ask, in light of recent scandals, did all of these reforms produce no significant or lasting improvements in our police?

Beyond the Media Images

Our hypothetical reasonable person would be misled by the scandals that have dominated the news, however. Quietly, and with little publicity, a number of police departments have made significant progress with regard to police accountability in recent years and have taken steps to curb excessive force, unjustified shootings, and other forms of misconduct. These efforts are the central theme of this book.

Ironically, some of the best indicators of this progress are to be found in the settlements secured by the Department of Justice in its pattern or practice suits. The reforms mandated in those agreements were not developed by attorneys in the Department of Justice, but were drawn from the policies and procedures already in place in the better police departments.[34] Local police departments developed, in an ad hoc fashion, comprehensive use of force reporting systems and early intervention systems—albeit under pressure from local civil rights and civil liberties advocates. The principle of an open and accessible citizen complaint system also developed locally in an erratic fashion in cities and counties across the country. Early intervention systems first appeared in a few departments almost 25 years ago. The significant development in the mid-1990s was that these various practices coalesced into a coherent program for enhancing police accountability.[35]

There is, in short, an enormous gap between the best and the worst police departments in this country. This phenomenon was noted several years ago by Herman Goldstein, arguably the premier authority of American policing over the past 40 years and the creator of problem-oriented policing. He observed that (at that time, the mid-1980s) the gap between the best and worst police departments was greater than at any time in American history.[36] The better departments had made a serious commitment to accountability and had the organizational capacity to explore new strategies and tools. The worst departments, meanwhile, were overwhelmed by the problems associated with the increase in drug abuse and violent crime. This book concurs with Goldstein's assessment and attempts to explain some of the reasons why.

A NEW FRAMEWORK FOR POLICE REFORM

The new police accountability consists of two elements. The first includes a set of specific strategies and tools designed to enhance accountability. The second is a conceptual framework that unites these strategies and tools into a coherent program of police reform.

New Strategies and Tools

If you quickly scan all of the settlements of Department of Justice pattern or practice suits you will quickly notice that they are very similar, with only slight variation from department to department. This similarity is due to the fact that they all embrace a short list of recognized best practices designed to enhance police accountability.[37] This list emerged slowly in the 1980s and 1990s, and only coalesced into a coherent package in the mid-1990s. These strategies and tools are, in some cases, entirely new devices and, in some other cases, simply extensions and elaborations of practices that have existed in policing for some time.

Use of Force and Other Critical Incident Reporting

The first element of the new police accountability involves written policies governing police use of force and other critical incidents, accompanied by the requirement that police officers file official reports each and every time they engage in one of these incidents. A *critical incident* is defined as any police action that poses a risk to the life, liberty, or dignity of a citizen. Additionally, these reports are automatically reviewed by supervisors to ensure that the officer complied with departmental policies. The overall strategy for holding officers accountable through written policies, required reports, and automatic review of those reports is discussed in detail in Chapter Three.

Open and Accessible Citizen Complaint Procedures

The second management tool for holding officers accountable is to maintain an open and accessible process for citizen complaints about officer conduct. Traditionally, police departments regarded citizen complaints as virtually hostile acts, to be fended off if at all possible. In the new accountability, however, citizen complaints are regarded as an important form of management information: that is, indicators of possible performance problems that need to be corrected. The basic principles and administrative details of open and accessible citizen complaint procedures, which can be maintained by either a police department or an external citizen oversight agency, are discussed in detail in Chapter Four.

Early Intervention Systems

The third management tool for enhancing accountability is an early intervention system that involves the systematic collection and analysis of data on officer performance for the purpose of identifying problems that need to be corrected. The computerized databases of EI systems utilize the use of force reports and citizen complaints that are generated by the first two tools discussed above. The nature and potential uses of EI systems are discussed in detail in Chapter Five.

External Citizen Oversight

The fourth tool for enhancing accountability is an external citizen oversight agency, which brings to a police department sustained input from experts who are not members of the police department. This book argues that the most effective form of citizen oversight is the relatively new concept of the police auditor. Unlike the traditional civilian review board, which reviews individual citizen complaints, the police auditor investigates the patterns and practices of a police department for the purpose of recommending policies that will correct problems that exist. The nature and functions of police auditors are discussed in Chapter Six.

A New Conceptual Framework

A tool is nothing more than that—an instrument that can be used properly or improperly, or not used at all. The new police accountability also involves a conceptual framework that unites and organizes the tools described above into a coherent instrument of police reform. The conceptual framework of the new accountability consists of several elements.

Changing Police Organizations

The basic goal of the new police accountability is organizational change. This represents a significant shift from a long-standing police reform emphasis on individual officers, or what is often called the rotten apple theory of police misconduct. The rotten apple theory persists and motivates many community activists because it has powerful emotional and political appeal. It personalizes misconduct and gives it a human face. Unfortunately, it is simplistic and ineffective. Most important, it does not address the underlying organizational and management causes of unjustified shootings and persistent use of excessive force. The new accountability thinks instead in terms of "rotten barrels," and directs its energies toward fixing the barrel. The changes involved, however, are decidedly lacking in emotional appeal: complex administrative procedures that have no human face, are difficult to implement and even harder to maintain over the long term, and whose results lie in the future rather than in the emotionally charged present. Firing a cop or a police chief has a certain cheap appeal, and chiefs can be rather easily dismissed. Far more difficult is the task of changing the culture of a police department, in the sense of developing informal norms of professional conduct and a habit of reporting and investigating misconduct.

In an initial assessment of the Justice Department's efforts under Section 14141, Debra Livingston argues that the "conclusion drawn by many police scholars" is that "police reform will be most effective ... when reform involves not simply adherence to rules in the face of punitive sanctions, but a change in the organizational values and systems to which both managers and line officers adhere."[38] The central argument of this book is that Livingston is absolutely correct—but that she only defines the starting point. The heart of the matter is how to change the organizational values of a police department and to maintain that change over time.

Controlling Street-level Officer Behavior

Changing a police organization ultimately is deeply intertwined with measures designed to control the behavior of individual officers on the street. To separate the two conceptually or in practice is to invite failure. After all, the millions of police–citizen interactions that occur every day are the ultimate test of whether the police are in fact accountable. The strategies and tools described in this book seek to reach deep down into police organizations and affect their day-to-day behavior. The organizational process is inseparable from the on-the-street product. Use of force reporting systems, although directed at individual officers, represent an organizational mechanism that defines new standards and new ways of doing business. Ultimately, or perhaps we should say hopefully, the strategies and tools of the new police accountability will alter the culture of the police organization.[39]

The Systematic Collection and Analysis of Data

Changing a police organization requires the systematic collection and analysis of data on officer performance. Critical incident reporting, an effective citizen complaint process, and an early intervention system are the specific tools for this purpose. The larger strategy is to develop a fact-based picture of officer activity for the purpose of identifying recurring problems that merit corrective action. The strategy of collecting and using systematic data for purposes of organizational improvement and improving the delivery of social services is increasingly used in other professions: medicine,[40] private enterprise, and other government agencies.[41]

One of the most celebrated reforms in policing in the past decade, in fact, is COMPSTAT, a program that collects and analyzes systematic data on patterns of crime and disorder for the purpose of focusing crime reduction efforts.[42] At the same time, systematic data collection embraces the principles of problem-oriented policing (POP), the first cousin of community policing and in many respects a more specific road map for action. The action framework for POP involves the SARA model of scanning, analysis, response, and assessment. Scanning, in this context, means the collection of systematic data; analysis involves the review of that data and the identification of problems that need attention; response is the action a department takes with regard to a problem; and assessment is the follow-up review on the impact of the response.[43]

The power of data is described by Merrick Bobb, Special Counsel to the Los Angeles Sheriff's Department (LASD). He concluded his report on the troubled Century Station by observing that "this chapter began with a discussion about numbers and ended with a discussion about management. This is how it should be."[44] The reform strategy of the new paradigm, instead of focusing on individual officers, uses comprehensive data about agency and officer performance to identify management problems that are likely to lead to misconduct by individual officers.

The Convergence of Internal and External Accountability

The tools and strategies of the new accountability involve a convergence of internal and external strategies for accountability that have historically been seen as competing alternatives. Traditionally, the police vigorously insisted that they have both the responsibility and the capacity to manage their own affairs—including matters of discipline—free of external intervention. Civil rights activists, despairing of the capacity of police departments to police themselves, have pursued a variety of external mechanisms of accountability.[45] These mechanisms have included the intervention of the courts, particularly with respect to constitutional standards for police work, and external citizen oversight agencies to handle citizen complaints. Figure 1.1 (p. 9) is a graphic representation of the competing accountability mechanisms.

The politics of police accountability over the past 40 years has to a great extent been defined by a bitter struggle between the claims of internal and external accountability, or between professional autonomy and external oversight. The focal point of this struggle has been the issue of civilian review boards. This book argues that the new police accountability involves a convergence of internal and external accountability and the emergence of a "mixed system" of mutually reinforcing accountability mechanism, as illustrated in Figure 1.1. The best example involves citizen complaint procedures. Traditionally, police departments regarded complaints as hostile acts on the part of citizens and did their best to reject or discredit them. In the new police accountability, an open and accessible citizen complaint process is an important source of management information. This information is entered into the departmental early intervention systems and analyzed to identify performance problems.

A similar convergence of internal and external mechanisms is embodied in police auditor systems (Chapter Six). The police auditor has emerged as an important form of citizen oversight, an alternative to the civilian review board that police departments traditionally bitterly opposed. The best-functioning police auditor systems bring an outsider's perspective to police operations and are designed primarily to promote internal police management reforms.

CONCLUSION: THE CHALLENGE OF CHANGING POLICE ORGANIZATIONS

Can Police Departments Change? Beyond Pessimism

If the focus of the new police accountability is to change police organizations, we have to confront the question of whether it is possible to transform them into organizations in which the commitment to accountability is self-sustaining. This is a major challenge. The history of police reform is filled with stories of highly publicized changes that promised much but evaporated over the long run with only minimal impact.

One of the more notable examples of the failure of accountability-related reforms would be the reforms developed by New York City Police Commissioner Patrick V. Murphy in the early 1970s. In the wake of the highly publicized corruption scandal investigated by the Knapp Commission (and generally associated with the name of officer Frank Serpico), Murphy decentralized corruption control in Field Investigative Units. These units were designed to be closer to streets where the problems existed than the old centralized unit, and therefore presumably more effective. Yet, as subsequent scandals and the 1994 Mollen Commission report revealed, these reforms had completely collapsed and blatant corruption flourished. Even worse, the Mollen Commission found a new and even more insidious form of corruption within the NYPD, a combination of brutality and graft.[46]

Many cynics believe that the American police are incapable of reforming themselves and that the police subculture is resistant to all efforts to achieve accountability. Regrettably, a review of police history lends an uncomfortable amount of support to this very pessimistic view.

Grounds for Optimism

The basic argument of this book is, to the contrary, that self-sustaining commitment to accountability is indeed possible. There is evidence of such a commitment in a number of law enforcement agencies across the country. As already noted, the elements of the new police accountability that the U.S. Department of Justice has incorporated into all of its consent decrees were already-existing programs in agencies around the country. Department of Justice litigators invented nothing new, but simply selected the best of these programs and packaged them into a coherent set of best practices.[47]

It is worth noting that in Los Angeles, Sheriff Lee Baca took the extraordinary step in 2001 of creating a *second* form of independent citizen oversight. The Office of Independent Review (OIR), led by a former U.S. Attorney and staffed with seven attorneys, duplicates much of the mission of the Special Counsel which had been in place since 1993.[48] Sheriff Baca could easily have not taken this step by claiming budget constraints and citing the existence of the Special Counsel. But he took it, making the LASD the first law enforcement agency in the United States to have two fully staffed forms of independent citizen oversight. (These two efforts are discussed in detail in Chapter Six.)

At the same time, the police departments in both Seattle and San Diego have undertaken programs of continuing review of use of force and racial profiling issues. Particularly notable is the fact that they are ongoing efforts to review policies and practices, include a high degree of citizen involvement, and produce reports that are readily available to the public on the web. Similar efforts are underway in other police departments.

It would be easy to overestimate the significance or the long-term prospects of these promising efforts. After all, the cycle of reform and failure has been repeated many times in the history of the American police. But there are also grounds for optimism. One of the central arguments of this book is that what is particularly new about the new police accountability is a more sophisticated understanding of the nature of the problem and a new set of tools and strategies to deal with it.

Grounds for Skepticism

The new police accountability is an exciting development. It holds great promise for the future. But we should not ignore the obvious problems it faces.

Candor requires that we emphasize the *promise* of the new accountability, as distinct from an achieved reality. By the prevailing standards of social science research there is only limited evidence that the tools and strategies described in this book in fact achieve their intended goals.

Even more disturbing, the evidence used in this book to argue for the new police accountability also includes many examples in which the new tools and strategies have not been properly implemented or have been allowed to fall into disrepair through administrative neglect. This evidence represents a substantial red flag about the prospects for meaningful and lasting reform. Will the new police accountability succeed? We cannot say at this point. It is too early in this national effort to draw any definitive conclusions about success or failure. What this book does do, however, is to map the landscape: to describe the new accountability mechanisms, explain in detail how they are intended to work, and carefully weigh the available evidence on successes and failures. Some years down the road, we will be able to say whether the effort succeeded, and if the book did identify the conditions of success, and if it did not at least help to tell us where we went wrong. In the meantime, more research is needed on which accountability mechanisms work and which ones work best.

TWO

THE ACCOMPLISHMENTS
AND LIMITATIONS OF
TRADITIONAL POLICE REFORMS

———— ⋅•◆•⋅ ————

To appreciate the significance of the new police accountability, it is necessary to place it in the context of past police reform efforts and what they accomplished and where they fell short. This chapter offers a brief critical review of the principal police reform efforts of the past.

Police reform is nothing new in the United States. Indeed, reform movements have been around since the first modern police departments were created in the 1830s. Throughout the nineteenth century, these efforts achieved little in the way of lasting improvement in policing. By the dawn of the twentieth century, the American police were still mired in corruption, brutality, and inefficiency, all the result of a pervasive system of political influence over police departments.[49] The new century brought the advent of the professionalization movement, the first police reform effort to achieve any lasting change. In the late 1950s and 1960s, civil rights and civil liberties activists turned to the courts as an instrument of police reform. Additionally, reformers over the years have sought to reform the police through political action, either in the form of special blue-ribbon commissions or permanent citizen oversight agencies. Law professor Barbara Armacost argues, consistent with the thesis of this book, that past police reform efforts have focused too much on "notorious incidents and misbehaving individuals" and not enough on the dysfunctional aspects of police organizations that sustain serious misconduct.[50]

The various police reform efforts can be usefully categorized into three groups that parallel the three branches of American government: administrative, judicial, and legislative. The first represents an internal strategy of reform, seeking change through the efforts of police administrators, whereas the latter two represent external strategies, seeking change through institutions outside police departments.

PAST POLICE REFORM STRATEGIES

The Administrative Strategy: Police Professionalization

The Police Professionalization Movement

The administrative strategy for achieving police accountability is embodied in the idea of police professionalism that emerged in the early years of the twentieth century.[51] The core principle of police professionalism is that law enforcement agencies have both a responsibility and a right to manage their own affairs, just as other professions enjoy a high degree of autonomy and control over their domains.[52] To this end, generations of police managers have strenuously fought the actual or threatened intrusions into their managerial prerogatives, whether by the U.S. Supreme Court, citizen oversight agencies, or police unions.[53]

The reform agenda of the professionalization movement included securing expert leadership for police departments, freedom from external (i.e., political) influence, the application of modern management principles to police organizations, and elevation of personnel standards for officers.[54] The agenda of the professionalization dominated police management thinking throughout the 1970s. Confidence in the basic tenets of professionalization began to crumble in the 1960s in the face of the civil rights movement and rising crime rates, and the community policing movement further challenged the professional model of policing in the 1980s. Nonetheless, despite these criticisms the accomplishments of the professionalization movement were significant and cannot be ignored.

By the 1960s, most police departments were far better managed than they had been 30 or 40 years earlier.[55] In terms of basic service delivery, most police departments deployed their officers on a rational basis; for example, they assigned more patrol officers to high-crime areas. This standard reflected a broader commitment to respond to all calls for service and respond as quickly as possible. Before the advent of the telephone, the two-way radio, and the patrol car, they could not as a practical matter receive and respond to calls for service. And as recently as the early 1940s, the departments in some major cities simply did not patrol certain areas at all. Personnel standards, in terms of minimum entry requirements and formal preservice training programs, were far higher than in the past (when there were basically no standards at all) and continued to rise, albeit at a slow pace. California and New York set a new standard in 1959 with laws requiring minimal preservice training for all sworn officers in the state. Compared with the utter lack of professionalism that prevailed throughout the late nineteenth century, when police departments were the "adjuncts" of political machines, the achievements of the professionalism movement were substantial.[56]

The Shortcomings of Professionalization

Despite its achievements, the professionalization movement still left many problems unaddressed, particularly with regard to accountability and the control of use of force and race discrimination. Two powerful external forces engulfed the police in the early 1960s and exposed these failings: the intervention of the U.S. Supreme Court into police operations and the civil rights movement. A later section of this chapter examines the impact of the Court on American policing. The civil rights movement exposed the lack of accountability in dramatic ways. Virtually all of the urban riots of the 1964–1968 period were sparked by an incident involving the police.[57] The police officer in the ghetto became the symbol of the national crisis in race relations. The principal demands of civil rights leaders became the hiring of more African American officers and the creation of civilian review boards to handle citizen complaints.[58]

One of the most devastating commentaries on police professionalization appeared in the 1968 Kerner Commission report on the riots that had swept the country in the previous four years. The report somberly observed that "many of the serious disturbances took place in cities whose police departments are among the best led, best organized, best trained and most professional in the country."[59] Although the comment was undoubtedly directed at the Los Angeles Police Department, which had a national reputation as the most professional department in the country, it served as a general indictment of the standards of police professionalism as well. In short, even the best departments had failed to serve the African American community properly, and in particular to control the use of both deadly and physical force.

An even more revealing index of the failure of professionalization is O.W. Wilson's textbook, *Police Administration*, which was generally regarded as the authoritative bible on the subject of how to manage a police department and which had guided almost two generations of police chiefs. A close reading of Wilson's text illuminates the extent to which professionalization simply ignored what we today regard as basic issues of police accountability. Even the 4th edition published in 1977 contains no reference to police discretion and no admission that it is pervasive in police work and can potentially result in serious mistreatment of citizens. It devotes a total of four pages (out of more than 600) to supervision through "written directives," a process that is now recognized as a basic element of police management (and is discussed at length in Chapter Three).[60] Moreover, the discussion is couched in vague and general terms, with no specific reference to the use of deadly force, physical force, high-speed pursuits, or other critical uses of police power. The typical police department manual was traditionally a small vest-pocket booklet, largely devoted to petty administrative rules about such things as grooming.

Wilson's primary focus was on the purely formal aspects of organization such as the proper organizational structure, the chain of command, and the rational allocation of patrol officers according to workload. Missing from this approach to police management was any explicit discussion of what police officers actually *do* on the street.[61] As the emerging social science literature on the police discovered, police officers routinely encounter complex and ambiguous situations, exercise broad discretion, and, in the absence of meaningful guidance, can engage in questionable actions.[62] Wilson's formalistic approach simply assumed that the proper organizational arrangements would necessarily result in the correct officer behavior. As we now know, and as this book argues in detail, the control of police power requires an array of detailed rules and procedures.

The Specific Failures

The failure of many police departments to establish basic standards of accountability, and in particular to control the use of force and curb race discrimination, is documented by the Department of Justice pattern and practice suits. The New Jersey State Police were engaging in gross race discrimination in traffic stops. The Pittsburgh Police Bureau had no meaningful system for controlling use of force or evaluating its personnel. The Rampart scandal exposed a special unit that was completely out of control in the Los Angeles Police Department. The Cincinnati Police Department consistently failed to address its conflicts with the African American community.

The following section lists some of the more notable failures and groups them into several general categories. Two aspects of this list are particularly notable. First, almost all involve a failure to implement the police profession's own standards (as opposed to standards developed by a human rights attorney, a federal judge, or an academic). Second, most of the examples that follow are of recent vintage, indicating a continuing failure in this regard in many large departments.

Failure to Adopt Well-Established
Principles for Patrol Management

- A 1987 investigation found that the Philadelphia Police Department had not redrawn its district boundaries in 16 years. Officers in the 35th district handled an average of 494 calls for service, whereas officers in the 5th district averaged 225.[63] In short, the third or fourth largest police department in the country went many years without examining whether it was using its officers in a rational and efficient manner.
- The Buffalo Police Department did not convert to the more efficient, and equally safe, practice of one-officer patrol cars until July 2003.[64] The greater efficiency and equal safety of one-officer patrol units was well established by the early 1960s. Yet Buffalo and many other departments (mainly on the East Coast) clung to this antiquated and inefficient practice into the twenty-first century.

Failure to Meet Established
Standards for On-the-Street Supervision

- In the late 1990s and early 2000s, investigations found several police departments that allowed the ratio of sergeants to officers on the street to exceed their own recommended standard of 1:8. The examples included the Century Station in the Los Angeles Sheriff's Department; the Riverside, California, Police Department; and the Philadelphia Police Department.[65] The police profession has long recognized that street-level supervision is the critical aspect of policing and developed a minimum standard, and yet these departments proceeded to violate that standard for extended periods of time.

- An observational study of the activities of street sergeants identified four different styles of supervision. One that Robin Engel labeled "supportive" supervision defined its role in terms of protecting officers from discipline by upper management.[66] In short, a common style of supervision is antithetical to principles of meaningful accountability. And it might be noted that the research was conducted in two departments implementing major community policing programs—that is, departments that were presumptively more progressive than many others.

Failure to Maintain Meaningful Personnel Evaluation Systems

- In the wake of the 1999 Rampart scandal, the Los Angeles Police Department's own internal Board of Inquiry Report concluded that the department's own personnel evaluation system was worthless. The report bluntly declared that "our personnel evaluations have little or no credibility at any level in the organization."[67] Nothing could be more basic to a professionally managed organization than an effective personnel evaluation system. And it should be noted that the LAPD is the department that long held the reputation as the most professional in the country.

- In the New York City Police Department, officer Michael Dowd's performance evaluation described him as having "excellent street knowledge" and said he could "easily become a role model for others to emulate." The 1994 Mollen Commission Report, however, cited Dowd as possibly the most corrupt and brutal police officer in the entire NYPD.[68]

- A 1997 report by the Police Executive Research Forum (PERF) concluded that "most performance evaluations currently used by police do not reflect the work officers do."[69] A 1976 report had reached a similar conclusion, suggesting that virtually no progress had been made in the intervening two decades.[70]

Failure to Respond to the High Costs of Civil Litigation

- In the 1990s, the City of Detroit paid out $124 million in lawsuits related to police misconduct. Despite an average cost of more than $10 million a year, neither the police department nor any other government official took any meaningful steps to correct the underlying problems and reduce this burden on the financially strapped city. Finally, the U.S. Department of Justice launched an investigation of the department and reached a settlement agreement in 2003 requiring a number of reforms related to officer use of force, the handling of persons in custody, and the supervision of officers.[71]
- Even though the concept of risk management was well established in private industry by the early 21st century, few police departments had formal risk management programs.[72]
- In the 1960s, the President's Crime Commission and other experts argued that police departments should employ legal advisors to provide expert advice that would prevent violations of constitutional law and misconduct that would lead to costly lawsuits.[73] Yet three decades later, the concept of legal advising remained a marginal feature of American police management.

Failure to Discipline Officers Found Guilty of Misconduct

- The Philadelphia Police Department did not discipline some officers against whom the department itself sustained a complaint. In other cases, not-guilty verdicts were rendered, or charges were withdrawn, with no or inadequate explanations, despite evidence indicating that such offenses did in fact occur.[74]
- In Schenectady, New York, the Department of Justice investigators found that officers routinely did not complete the required use of force reports, and that completed reports were often not reviewed. The basic idea of controlling use of force through a report and review process has been established for more than thirty years (see Chapter Three). Yet some departments have not adopted it.

Discussion

The significance of the above list is that it represents serious failures to implement the law enforcement profession's own standards. These are not standards developed by outsiders. The nature of these failures prompts the question of how they were allowed to occur. The obvious answer is that the top leadership in each of these departments failed to "mind the store," as it were, and to ensure that these basic standards were being met. This answer is not completely adequate, however, and we need to inquire into the nature of this failure. Part of the answer lies in the very nature of complex bureaucracies. Failure to ensure full compliance with all written policies and procedures is probably endemic in private corporations, hospitals, universities, and so on.[75]

Another part of the answer lies in the crisis management nature of American policing. As a matter of course, the leaders of every large police department are buffeted by emergencies that require their attention: a particularly vicious crime; public demands for more police protection; a shooting incident; an allegation of excessive force; a rumor about some other embarrassing officer misconduct; a budget crisis; and so on. In this atmosphere it is easy to neglect routine procedures and lose sight of larger goals. Personnel shortages are especially pernicious. With this in mind, it is easy to imagine how departments fail to maintain the recommended sergeant–officer supervisory ratio. They find themselves without enough sergeants on a particular shift and, in response, simply double up by assigning extra patrol officers to those sergeants on duty. What begins as a temporary solution gradually becomes a permanent way of operating.

The failure to adopt meaningful personnel evaluations, meanwhile, is probably symptomatic of the pervasive "grade inflation" in American society in which everyone is "above average." A similar problem with personnel evaluations exists in the military. Meaningful personnel assessment is difficult in any organizational context. Rather than risk conflict and bad feelings, it is simply easier to give everyone high marks. It is especially difficult in human service organizations, such as policing and teaching, in which there are no quantifiable outcome measures such as sales or profits.

In short, some of the problems afflicting police departments are not unique to the police, but are endemic in modern bureaucracies and larger trends in American culture.

What, then, is the solution to this problem? Clearly, police departments need some kind of regular system of inspections similar to the process used in the military. Although some departments have nominal inspections units, they are not found in all departments, and where they exist they are not sufficiently independent of the organizational command structure to be critical of existing operations. The current law enforcement accreditation standards require a system of Inspectional Services.[76] The new police accountability incorporates a potential solution to this problem. Chapter Six describes the emerging practice of the police auditor as a permanent external citizen oversight agency that has the authority to audit, monitor, and inspect any and all aspects of police operations.

A Note on Accreditation of Law Enforcement Agencies

The shortcomings of professionalization are also evident in the law enforcement accreditation process administered by the Commission on Accreditation for Law Enforcement Agencies (CALEA).[77] Professional self-regulation through accreditation is one of the hallmarks of all established professions. Accreditation in policing, however, was very late in developing and today has had only a limited impact on policing. By early 2004, only about 500 of the nearly 18,000 law enforcement agencies in the country were accredited. Accreditation is entirely voluntary and there is absolutely no penalty, such as the loss of federal funds, for a police department that is not accredited.

The current accreditation standards themselves are also cause for concern. With rare exception they do not embody substantive requirements—what might be called standards of care—on specific issues. Typically they require that an agency "have a written policy on" or that a particular staff member be responsible for a particular issue. They do not, however, specify what that policy should be. Standard 41.2.2 on vehicle pursuits, for example, requires that the department have a written directive on the subject and that it cover "evaluating the circumstances" of a potential pursuit.[78] But it provides no substantive guidance on what circumstances to consider (e.g., road conditions) or how to weigh conditions if faced with a pursuit decision (e.g., balancing bad roads against the dangerousness of the fleeing suspect). The standard on in-service training (Standard 35.1.1) requires that all sworn officers receive in-service training annually. But it does not specify how many hours of training they should receive, nor does it mandate coverage of particular subjects, such as human relations, communication skills, traffic stops, the use of informants, and so on.[79]

With respect to the investigation of alleged officer misconduct—a key element of accountability—the CALEA Standards do not specify any details about a department's Internal Affairs or Professional Standards unit. The CALEA Standards require that a department have such a unit but do not specify how many investigators it should have relative to the size of the department or how they should be selected and trained.

Only three CALEA Standards prescribe a substantive standard of care, and only one of those is very specific. The Standard on use of deadly force (Standard 1.3.2) embodies the prevailing defense of life standard. With regard to the recruitment of racial and ethnic minority and female officers, the CALEA Standards (Standard 31.2) state that the composition of an agency should reflect the composition of the community served. This is consistent with the standard recommended by other experts and is roughly comparable to the standard used by the courts in affirmative action cases. The existing American Correctional Association (ACA) Standards, by contrast, are very specific on these and other issues, providing a definite standard of care that must be met.[80]

The Judicial Strategy:
The Courts as an Instrument of Police Reform

The failure of police professionalization to address critical issues of police use of force and race discrimination led civil rights activists to turn to the courts as an instrument of police reform.[81] The resulting judicial strategy of reform consisted of three different avenues for pursuing higher standards of police accountability: constitutional law, tort litigation, and criminal prosecution.

The Constitution as a Code of Criminal Procedure

The U.S. Supreme Court emerged as a potent force for police reform during the 1960s, and its impact continues today, if only indirectly. In a series of highly publicized decisions that are among the most famous in the history of the Warren Court, which issued many controversial decisions, the Court intervened in previously hidden matters of routine police work and imposed new standards of conduct based on principles of constitutional law. The Court's decisions relating to the police were only one part of a larger due process revolution affecting the entire criminal justice system, which in turn was only one part of the broad impact of the Warren Court on American law and life.[82] The Court's most famous decisions on policing, *Mapp v. Ohio* (1961) and *Miranda v. Arizona* (1966), are well known, and in fact have permeated popular culture. Virtually every schoolchild today knows that the police must advise a criminal suspect of his or her right to remain silent.

The impact of these decisions was profound and multifaceted. At the basic level, the Court stepped into important aspects of police operations and, finding violations of constitutional protections, drafted a set of rules that included meaningful penalties. Most important in terms of its overall impact, the Court defined national standards for certain aspects of police operations. This was a particularly profound innovation in the context of the highly decentralized structure of American policing, with nearly 18,000 independent state and local agencies and no process for establishing much less enforcing national standards.[83] To an even greater extent, the federal courts fashioned national standards for correctional institutions between the late 1960s and the mid-1980s.[84] In a broader sense, the Court's decision threw a spotlight on previously hidden aspects of police work, illuminating routine practices and the problems associated with them. This not only educated the public about police procedures but provided that education in the context of constitutional standards that should govern policing.

Scholars have expended much energy attempting to assess the impact of both the *Mapp* and *Miranda* decisions, attempting, for example, to measure the number of cases "lost" as a result of constitutional law violations.[85] These narrowly focused efforts have missed the larger effects of the Court's role as a watchdog of the police. In addition to the impact on public understanding, the intervention of the Court was a profound shock to American policing and set in motion a number of reforms, most of which continue to the present day. Faced with new rules and penalties, the police scrambled to improve both their personnel and their procedures. Efforts to improve recruitment and training standards received a major boost, as departments realized they needed officers who could understand and comply with detailed court mandates. Similarly, departments undertook new efforts to provide officers with guidance on how to comply with court decisions, either through in-service training or direct guidance from local prosecutors.[86] Departments provided officers with written guidance on how to comply with court decisions, and from this effort developed both the modern police department standard operating procedure manual and the general strategy of directing officers through administrative rulemaking.[87] Chapter Three examines the important development of this process with regard to officer use of force.

The accreditation movement, for all its limitations, was a direct outgrowth of the intervention of the courts as the law enforcement profession sought to preempt court intervention through self-regulation. A similar process occurred in corrections, as correctional officials saw accreditation as a way of protecting themselves against the flood of prisoner's rights litigation that began in the early 1970s.[88]

All of these very positive developments have continued to the present day, and continued long after the Supreme Court began to withdraw from an activist role as watchdog of the police in the 1970s. By establishing constitutional principles as a minimum standard for police work, the Court reshaped the debate over police reform and stimulated lasting reform efforts.[89]

The Limits of the Supreme Court as a Mechanism for Police Accountability

During the heyday of the Warren Court, civil rights and civil liberties advocates invested enormous hopes in the Court as an instrument of police reform. And inevitably, they have been extremely dismayed at the subsequent withdrawal of the Court from an activist role as watchdog of the police. Even at the height of the Court's activism, however, a number of perceptive experts recognized the limits of the Court as an instrument of police reform. This included observers who fully supported the intent and the result of the Court's activist role.[90]

The principal limit on the Court is that it lacks the institutional capacity to ensure compliance with its own decisions on a day-to-day basis. Although it is true that both *Mapp* and *Miranda* include significant penalties for violating constitutional protections, all experts in policing recognize that police work includes abundant opportunities for evading them. In the immediate aftermath of *Mapp*, for example, experts worried that officers would simply lie and claim that suspects "dropped" the seized contraband.[91] An observational study of Miranda found not only that a high percentage of suspects voluntarily confessed but that in about 30 percent of all cases, the interrogating officers lied to them.[92]

Even more serious is the fact that so many critical aspects of routine policing fall outside the purview of any court decision defining constitutional standards. For example, although police experts recognize that a proper ratio of sergeants to patrol officers is a key element of good supervision, it is doubtful whether failure to meet that standard rises to the level of a constitutional law issue. One of the central arguments of Chapters Three and Four of this book is that both the effective control of police use of force and the handling of citizen complaints involve innumerable operational details that go far beyond anything that even a very activist Supreme Court is likely to address. The American Bar Association *Standards Relating to the Urban Police Function* made this argument 30 years ago and it is still highly relevant today.[93]

Suing the Cops: Reform Through Civil Litigation

Civil rights and civil liberties activists have also used tort litigation, under federal or state law, as a strategy for enhancing police accountability. Apart from compensation for individual plaintiffs, the reform strategy assumes that if the dollar cost of police misconduct is raised to a critical level, local elected officials will respond and, by adopting the techniques of risk management, impose meaningful police reforms.[94] The reform assumption underlying this strategy is the belief that increasing the financial pressure on cities and counties might succeed where appeals to constitutional law and human rights did not.

Although there have been numerous small victories, there is little evidence that civil litigation has been successful as a general strategy for reforming the police. Academic studies of the strategy have generally found little direct impact on police reform.[95] The example of Detroit cited above—where the city was paying out over $10 million a year for 10 years with no effort to improve the police—is probably an extreme case, but not an entirely isolated one. Los Angeles, New York City, and other cities and counties have also paid out large sums for police misconduct over the course of many years.[96] To be sure, many suits have forced local departments to adopt a new or revised policy on a particular aspect of police operations, but until the advent of the Department of Justice "pattern or practice" suits in the past decade this approach has not succeeded in achieving comprehensive reform in any department. (It may be that employment discrimination litigation, involving both race and gender, has been more successful in eliminating bad practices than is the case with use of force and other accountability issues. Confirming or refuting this impression is beyond the scope of this book, however.)

The flaw in the civil litigation strategy is the assumption that public officials will necessarily act in a rational and coordinated manner in response to a problem. Instead, the response has been more often one of indifference and disconnection. Human Rights Watch, for example, quoted one police internal affairs officer as saying "civil cases are not our problem."[97] Barbara Armacost argues that "many police departments apparently consider the money they pay out in damages and settlements as simply a 'cost of doing business,'" and quotes former LAPD Police Chief Daryl Gates to that effect.[98] Generally, one agency of government (the police) perpetrates the harm, another agency defends it in court (the law department), and a third agency writes the check (the treasurer). Absent from this scenario is a coordinated risk management program that seeks to identify the sources of litigation costs and then correct the underlying problems. Why several generations of mayors have failed in this regard is both an intriguing question and a sad commentary on municipal government in America. A recent survey of police risk management programs, moreover, found that even those programs that claim to have reduced litigation costs do not collect and publish data that would verify these claims.[99]

One major exception to the observations in the previous paragraph is Los Angeles County, which created the Special Counsel to the Los Angeles County Sheriff's Department (LASD) in 1993 in response to high civil litigations costs.[100] But this is a new development and one closely associated with the new police accountability (see Chapter Six).

Federal Pattern or Practice Litigation

Civil litigation emerged as a potent strategy for police reform in the 1990s. Section 14141 of the 1994 Violent Crime Control Act authorizes the U.S. Department of Justice to bring civil suits against law enforcement agencies where there is a "pattern or practice" of abuse of citizens' rights. Unlike the civil litigation described above, the law specifically authorizes the Department of Justice to sue organizations (rather than individual officers) for the purpose of bringing about organizational change. By 2004, the Department of Justice had reached settlements with 19 law enforcement agencies, either through consent decrees, memoranda of understanding, or settlement letters.[101]

As discussed in Chapter One, the various settlements require, with only slight variation, a set of reforms that comprise the new police accountability: improved use of force reporting and investigation of force incidents, improved citizen complaint procedures, and an early intervention system. Most also require the collection of data on traffic or pedestrian stops in response to allegations of racial profiling. Each of these reforms is discussed in detail in the next three chapters of this book, and those discussions will not be repeated here.

It is worth noting that litigation under Section 14141 represents an approach to systemic organizational reform for law enforcement agencies that the federal courts began with respect to prisons in the late 1960s. The prisoners' rights movement effected sweeping changes in prison practices, eliminating many unjust practices, revolutionizing the authority structure within prisons, and setting in motion an accreditation process that far exceeds law enforcement accreditation in its scope.[102]

The relevant question is why did the federal courts ignore police organizations while ordering sweeping reforms in prisons? Herman Goldstein argues that the critical turning point was the Court's 1976 decision in *Rizzo v. Goode,* in which it declined to order administrative reforms in the Philadelphia Police Department.[103] The District Court had heard what the Supreme Court termed a "staggering amount of evidence," including 250 witnesses over a 21-day period. Yet the Court held that none of the plaintiffs could show that they had been harmed by the police practices cited and, further, that none of the city officials named as defendants were alleged to have "acted affirmatively" in depriving the plaintiffs of constitutional rights. The Court declined to accept the role of "fashioning of prophylactic procedures" designed to minimize employee misconduct. Why the Court declined to address deficiencies in police policy and practices while at the same time choosing to address similar kinds of failures in prisons is a question that is beyond the scope of this book but one that experts on the Court can address. One possible explanation is the deeply ingrained deference to the police. After all, the federal courts were then in the business of ordering all manner of administrative reforms in prisons designed to minimize misconduct, and to appoint special masters to oversee their orders.[104] One can only speculate about how the history of American policing might have been different had the Supreme Court decided *Rizzo* differently in 1976.

Criminal Prosecution

Community activists have also sought to curb police misconduct through criminal prosecution of officers guilty of criminal acts related to excessive physical force or unjustified shootings. As a strategy for reforming the police, this approach assumes that the successful conviction of officers will not only remove bad officers from the police department but also deter future misconduct by other officers.

Criminal prosecution, however, has been an extremely weak instrument of reform.[105] Convictions of police officers are extremely difficult to obtain. Local prosecutors, by the very nature of their role, have very close working relationships with local police departments and are reluctant to file criminal charges. The resources of the U.S. Department of Justice, meanwhile, are extremely limited, given its many responsibilities and the fact that there are more than 18,000 local law enforcement agencies in the United States. The legal standard of proof beyond a reasonable doubt that the officer had criminal intent is extremely difficult to meet. Officers can always claim that they faced a threat to their own lives and were therefore justified in using force. Judges and juries are extremely deferential to these claims and to the police in general. Even though there have been successful prosecutions of officers, they do not appear to have any deterrent effect. In both New York City and Philadelphia, for example, many officers have been convicted over the past three decades, yet both departments have been beset by repeated scandals involving both corruption and brutality.

Summary

The judicial strategy for reforming the police has had a very mixed record of success. Criminal prosecution of officers has proven completely inadequate, and tort litigation has had only limited success. Reforming the police through constitutional law, however, had a transformative role in reforming the police in the 1960s. In addition to establishing the principle that the police should be held accountable to standards of constitutional law, it stimulated a wide range of reforms that continue to this day. Although the Supreme Court began to withdraw from an activist role toward the police in the 1970s, federal district courts are now assuming a very important role regarding police accountability through federal suits authorized under Section 14141. The substantive issues in the settlement of those suits form the basis for Chapters Three, Four, and Five of this book.

The Legislative Strategy: External Oversight of the Police

Community activists have also turned to the political arena in response to police misconduct, demanding external oversight of the police. In a familiar process known as "scandal and reform,"[106] the exposure of police abuse (corruption, excessive force, etc.) mobilizes public opinion and forces elected officials to take some kind of action. External oversight has taken two different forms: one-time, blue-ribbon commissions and permanent external oversight agencies to handle citizen complaints against police officers.

Blue-Ribbon Commissions

Blue-ribbon commissions are a familiar feature of the American political landscape. In response to a perceived social problem, chief executives at the local, state, and national levels typically appoint a panel of experts to investigate the problem and prepare a report with a set of policy recommendations. There is a long history of "riot commissions" appointed in response to episodes of urban racial violence,[107] to exposés of police corruption, and to incidents of excessive force. The Christopher Commission (1991), appointed in the wake of the Rodney King beating in Los Angeles, is simply the best-known recent example.[108]

National-level blue-ribbon commissions have made important contributions to American policing. The Wickersham Commission (1931),[109] the President's Crime Commission (1967),[110] and the American Bar Association *Standards Relating to the Urban Police Function* (1974)[111] documented existing problems, defined minimum standards, and made specific recommendations that strengthened the hand of reformers in local communities.

Blue-ribbon commissions suffer from one inherent and serious weakness, however. By their very nature they lack any capacity to implement their own recommendations and ensure that reform occurs.[112] Commissions are temporary bodies that disband once their final report is released. (The reports do become useful sources of information for academic studies of the police, however.) Implementation depends on a voluntary effort by a police department itself. In some instances, the original scandal results in the appointment of a new police chief who makes a sincere effort to implement the recommended reforms. The typical result is that the political momentum for reform wanes as the original crisis fades into memory and public attention (particularly the attention of the news media) moves on to new crises.[113]

Chapter Six of this book argues that police auditors, a new form of citizen oversight that has appeared in the past decade, overcome the inherent limitations of blue-ribbon commissions. Most important, as permanent agencies they have the capacity to monitor implementation of their recommendations.

Citizen Oversight Agencies

A second legislative strategy for curbing police misconduct involves creating a permanent external oversight agency to handle citizen complaints. In the police–community relations crisis of the 1960s, the creation of "civilian review boards" became one of the principal demands of civil rights groups.[114] (This book uses the term *citizen oversight* because it is more representative of the variety of external agencies and procedures that have developed in recent years.)[115] The civilian review movement suffered seemingly fatal blows in the late 1960s with the demise of review boards in both New York City and Philadelphia.[116] It quietly revived in the early 1970s, however, and by the mid-1980s, citizen oversight, including important new forms, was an established part of American policing. By 2004, virtually all of the big city police departments in the United States were subject to some form of citizen oversight, with new agencies being created at a steady rate.[117]

Citizen review of complaints is based on the assumption that police departments are inherently unable to police themselves, as a result of both bureaucratic self-interest and the power of the police subculture, and that an external citizen oversight agency, staffed by people who are not sworn officers, will necessarily conduct more independent, thorough, and fair investigations of citizen complaints. However, the spread of citizen review boards has left many of its advocates with a bitter taste. The New York Civil Liberties Union, for decades the leading advocate of citizen oversight in New York City, has also been the leading critic of its child, the Civilian Complaint Review Board (CCRB), issuing a series of reports on its shortcomings.[118] A recent report of a New Orleans Police–Civilian Review Task Force was highly critical of the Office of Municipal Investigations (OMI) that has been in place since 1981.[119] An investigation of the Detroit Police Department found serious accountability problems, despite the existence of a civilian-staffed office of citizen complaints since 1973.[120]

Civilian review agencies have often lacked the authority to accomplish their stated objectives: for example, promising "independent" review of citizen complaints without the power to conduct any investigations.[121] Others have had the power to conduct independent investigations but have lacked the necessary staff and budgetary resources.[122] Some have suffered from poor management.[123] Others have failed because of a lack of political support, disinterest by police management, or staunch opposition from the local police union.[124]

An increasing number of observers argue that the traditional strategy of civilian review of complaints—to the extent that it is limited to reviewing individual citizen complaints—is not likely to reduce police misconduct or produce long-term improvements in the quality of police services even in the best of circumstances. First, the vast majority of citizen complaints are "swearing contests," with no independent witnesses or evidence to support either side.[125] This explains why citizen review agencies generally do not sustain a significantly higher rate of complaints than internal police complaint procedures.[126] Second, even if a review agency did sustain a significant number of complaints, there is no persuasive evidence that this would deter officer misconduct. Finally, and perhaps most important, focusing on individual complaints tends to make rank-and-file officers scapegoats for police misconduct when such misconduct is the product of an organizational culture that permits it to exist.

Recognizing the inherent limits to sustaining citizen complaints, a number of experts on oversight argue that oversight agencies should focus on reviewing policies and procedures for the purpose of changing the organization and preventing future misconduct. That conclusion informs this book and places in context the role of police auditors as an alternative form of citizen oversight (see Chapter Six). Indirectly, it is also part of the rationale for early intervention systems (Chapter Five). Given the problems in sustaining individual citizen complaints and also of obtaining criminal convictions in extreme cases, it may be much easier to spot officer performance problems at an early stage and correct them before a major incident occurs.

CONCLUSION: THE LESSONS OF THE PAST

The new police accountability builds on the lessons of these past reform efforts. The following section attempts to identify the most important of those lessons.

First, many past reforms were superficial or purely formalistic and failed to change day-to-day police work on the street. One popular reform illustrates the point. For the past 100 years, reformers have urged higher educational requirements for police officers. The argument here, however, is that in a poorly managed department, with no meaningful standards of accountability, even the better educated officers will sink to a low standard of performance.

Second and closely related to the first point, many reforms designed to achieve greater accountability failed to reach deep into the police organization and affect police officer behavior on the street. The most notable example would be Supreme Court decisions on such issues as searches and seizures or interrogations. As many critics have pointed out, the Court has no mechanism for enforcing its decisions (apart from ruling in a subsequent appeal in a new case). Ultimately, it falls to police departments to develop the internal procedures for ensuring officer compliance with Supreme Court decisions and other legal mandates.

Third, past reform efforts largely neglected the role of supervisors in controlling the critical actions by officers on the street. In all of the literature on policing there is an alarming absence of information about street-level sergeants.[127] This is particularly surprising because all police commanders and most outside experts agree that the sergeant is the key to day-to-day policing. Yet we have precious little information about what sergeants do, how they think about their role, their activities that are most effective in improving the performance of officers under their command, the nature and quality of their training, variations in sergeants' activities across departments, and so on.[128]

Fourth, piecemeal reform, with no overarching strategy for organizational reform, failed to appreciate the interconnectedness of various reforms. It does little good to improve police training, for example, if officers know that the department tolerates use of excessive force on a routine basis. Reversing the order changes the role of training dramatically. If a department makes a strong effort to control the use of force, and officers see this happening, training over the use of force becomes very meaningful. Along the same lines, a department can adopt a state-of-the-art policy on use of force, but if sergeants do not effectively supervise and review force reports as intended, the written policies become empty bureaucratic gestures.

Finally, past reform efforts never developed institutionalized procedures for sustaining reform over time. There were no procedures for monitoring reforms that were instituted and for ensuring that they are implemented and that the reform process continues over time. The classic illustration of this point, discussed in this chapter and elsewhere, is the traditional blue-ribbon commission.[129] The standard response to a major police problem—a corruption scandal, a racial disturbance—has been to appoint a special commission that investigates and then issues a report containing a set of recommended reforms. Such commissions, however, have lacked the power both to implement their own recommendations and to follow up on whatever changes were made in the police department. In addition, there are numerous examples of reform chiefs appointed in the wake of a scandal or other problem who did make a number of important changes. Too often, however, subsequent events revealed that the changes were never fully implemented in the first place or were allowed to wither over time.

In a variety of ways, elements of the new police accountability address these historic shortcomings in police reform. The question, of course, is whether they will succeed where other efforts have failed. That question is discussed at length in the final chapter of this book.

THREE

USE OF FORCE REPORTING

—————•◦•———

ACCOUNTING FOR ONE'S ACTIONS

The core operating principle of the new police accountability is that police officers are required to account for their behavior. In practice, this means that police departments have a *written policy* clearly specifying when use of force is appropriate, require officers to complete a *written report* after each use of force incident, and have each report *reviewed* by supervisors. The report and review process is now a recognized best practice in policing. The Department of Justice report *Principles for Promoting Police Integrity* recommends that "Agencies should develop use of force policies that address use of firearms and other weapons and particular use of force issues such as: firing at moving vehicles, verbal warnings, positional asphyxia, bar arm restraints, and the use of chemical agents."[130] This chapter describes how the new police accountability extends the report and review requirement to a broader range of critical incidents involving the use of police powers.

The idea that officers should be subject to detailed rules and have to account for their actions in writing is actually relatively new in American policing. The absence of meaningful controls over the use of police powers in the past is, by today's standards, truly astonishing. As recently as the early 1970s in many departments police officers were sent out onto crime-ridden streets, armed with deadly weapons and trained in *how* to fire those weapons, but with absolutely no guidance on *when* to fire those weapons. A 1961 survey found that about half of the departments surveyed relied on an "oral policy."[131] The 1963 edition of O.W. Wilson's influential textbook *Police Administration* said nothing about the use of deadly force.[132] A recent report by the Philadelphia Police Department's Integrity and Accountability Office quoted officers who recalled the 1970s as the "wild west," where it was "open season" and a "free for all." Warning shots and shots at fleeing suspects (two actions now prohibited by all departments) "occurred with alarming frequency."[133]

In addition to the lack of written policies, prior to the 1970s most police officers did not have to complete detailed reports about use of force incidents. Even in those departments in which some kind of formal reports was required, supervisors generally did not conduct rigorous reviews of those reports with an eye toward disciplining officers who violated policy. The presumption was that officers used good judgment and should not be second-guessed in dangerous encounters. In this sense, they were literally unaccountable for their behavior.[134]

Holding police officers accountable by requiring them to explain their use of force has important collateral benefits. The principal source of police–community relations tensions has been the deeply held belief in the African American community that police officers can shoot to kill and use excessive force with impunity. As the court-appointed monitor in Los Angeles points out, fair and impartial investigations of use of force incidents not only ensure accountability for individual officers but are crucial to maintaining "the community's faith in the system."[135]

THE ORIGINS OF USE OF FORCE POLICY

Meaningful controls over police use of force began with an effort to control police shootings. It is hardly surprising that they began with the use of deadly force. Taking a person's life is the ultimate use of police authority—and the police are unique among social institutions in possessing this power. Additionally, fatal shootings by police have long been the most volatile civil rights issue. The killing of an African American man by white authorities has enormously powerful symbolic resonance, evoking images of lynchings and the still volatile issue of capital punishment. Several of the riots of the 1960s were sparked by the fatal shooting of an African American by a white police officer.[136] Data from the 1960s and early 1970s indicated a shocking disparity of eight African Americans shot and killed for every one white person.[137] In Memphis, Tennessee, between 1969 and 1974 officers shot and killed 13 African

Americans and only one white person in the "unarmed and not assaultive" category.[138] Shooting related crises have continued to the present day. In April 2001, the 15th fatal shooting of an African American in 5 years by the Cincinnati police precipitated a 1960s-style riot marked by property destruction, a curfew, and mobilization of the Ohio National Guard.

The New Deadly Force Policy in New York City, 1972

The historic turning point on use of force reporting occurred in 1972 with a new deadly force policy developed by New York City Police Commissioner Patrick V. Murphy. There may well have been precursors to Murphy's policy, but they have been lost to history and, in any event, had no meaningful influence on national police practices.[139] In part because the NYPD is the largest department in the country, its new deadly force policy had an enormous national influence. Additionally, it was subject to a rigorous evaluation by Professor James Fyfe, and his finding that the policy reduced shootings lent important academic support to the new approach.[140]

The 1972 NYPD policy included the two elements that have formed the basis of all use of force policies over the past 30 years. Substantively, the policy confined discretion by clearly specifying when force can be used and when it is not appropriate, replacing the very permissive *fleeing felon rule* with the restrictive *defense of life* rule. Officers are permitted to use deadly force for the protection of their own lives or the lives of other people. In addition, the policy prohibited firing a weapon for a number of specific purposes, including warning shots, shots intended to wound a suspect, and shots at or from moving vehicles.[141]

Procedurally, the NYPD policy required officers to complete a written report after each firearms discharge and mandated an automatic review of each report by supervisors. The review was conducted by a Firearms Discharge Review Board composed of several high-ranking commanders. Other departments have developed different procedures for these reviews, but the basic principle of an automatic review of each incident has become standard practice.

The basic elements of the NYPD policy soon won favor among police experts. A 1977 report on deadly force by the Police Foundation endorsed the mandatory reporting and automatic review requirements.[142] In 1981 the U.S. Civil Rights Commission's influential report, *Who is Guarding the Guardians,* recommended that "Unnecessary police use of excessive or deadly force could be curtailed by . . . strict procedures for reporting firearms discharges."[143] By the time of the 1985 Supreme Court decision in *Tennessee v. Garner* limiting police shootings under the Fourth Amendment, most big city police departments had already adopted deadly force policies that were far more detailed and restrictive than the Court's decision, reflecting a national consensus on the basic elements of the original 1972 NYPD policy.[144]

The Impact of Administrative Controls

Empirical research indicated that the new controls over police shootings had a positive effect. James Fyfe's analysis found that the 1972 NYPD rules reduced total firearms discharges by 30% over the next 3 years. National data, meanwhile, indicate a significant reduction in the number of citizens shot and killed by the police each year from its peak in the early 1970s to the 1980s (at which point the number has fluctuated). Even more important, the racial disparity in persons shot and killed has narrowed from a ratio of 6 or 8 African Americans for every white person shot in the mid-1970s to a ratio of 3 to 1 by the late 1990s.[145] In Memphis, where the old fleeing felon rule had resulted in 13 African Americans and only 1 white person shot and killed in the "unarmed and nonassaultive" category, the new restrictive policy resulted in no fatal shootings of any people, white or African American, in this category by the late 1980s. In short, the defense of life rule not only achieved its intended goal of eliminating fleeing felon shootings but in the process reduced the worst racial disparities.[146]

The Development of Critical Incident Policies

Since the initial breakthrough in the early 1970s, use of force policies have developed in four important directions. First, the use of written policies has been extended to cover a steadily increasing range of police actions, including use of physical force, high-speed vehicle pursuits, the use of canine units, and other actions. The emerging standard is that all critical incidents should be covered by a written policy requiring a report by the officer and an automatic review by supervisors. Critical incidents are defined here as any police action that has a potentially adverse effect on the life, liberty, or dignity of a citizen.

Second, the content of use of force policies has become increasingly detailed, covering more potential situations within a general category. With respect to the use of physical force, for example, can an officer kick a suspect?

If so, where? If so, with the foot, or knee, or both? Under what circumstances? This is not a hypothetical issue, because some departments do authorize "distraction" techniques that can involve kicking. It is increasingly recognized that to effectively control the use of force and to avoid any ambiguity, use of force policies need to address all possible applications.

Third, increased attention has been given to the nature of the review of use of force reports. Experts increasingly recognize that it is not sufficient merely to have a written policy requiring "a review." As is explained in detail later in this chapter, the emerging standard is to have specific policies requiring commanders to respond immediately (e.g., "roll out") to serious force incidents, to require investigators to canvass the scene for potential witnesses, not giving an automatic preference for the statements of the officer being investigated, and so on. And as Chapter Four explains, similarly detailed requirements have developed for the investigation of citizen complaints. In this regard, it is noteworthy that the consent decrees negotiated by the U.S. Department of Justice have become longer and more detailed, a trend that reflects a growing sophistication about what is required for an effective use of force policy.[147]

Also, it is no longer acceptable that the review of an incident be limited to the question of whether an officer violated department policy or committed a criminal offense. The emerging standard is that the review should inquire into whether an incident raises policy or training issues that the department needs to address.[148] In short, individual incidents should not be treated as isolated events but should become an occasion for organizational self-scrutiny and change.

Fourth, reports on critical incidents are increasingly subject to aggregate analysis for the purpose of identifying patterns of officer behavior that merit closer attention by the department. The new tool for such analyses, early intervention systems, is discussed in detail in Chapter Five.

Uneven Progress

Despite the emergence of a general consensus on use of force policy through the 1980s and 1990s, it would be a mistake to assume that all departments have been in step with this trend. On the contrary, as investigations by the U.S. Department of Justice have revealed, many police departments failed to adopt the new standards. The 1997 suit against the Pittsburgh Police Bureau found that the department's use of force policy did not meet national standards. The 2003 Department of Justice investigation of the Schenectady Police

Department found that its policy did "not limit the use of deadly force to situations involving an imminent threat to the life of the officer or another person. In fact, the policy appears to state that the use of deadly force may be justified even when there is no imminent threat to the life of the officer or another person." In addition, the Schenectady policy did not "adequately identify types of force that constitute deadly force."[149]

THE ADMINISTRATIVE RULEMAKING MODEL

The general model for police use of force policy is derived from the field of administrative law, and in particular the work of Kenneth C. Davis. His short 1975 book on *Police Discretion* was the first full discussion of discretion in policing, and it provided a brief description of how the administrative rule-making process, which was then well developed in other areas of government, could and should be applied to policing.[150]

The Framework: Confine, Structure, and Check Discretion

Davis's administrative rulemaking approach to the control of discretion, which he had earlier set forth in his book *Discretionary Justice,* involves a three-stage framework of confining, structuring, and checking discretion.[151]

Confining Discretion

Confining discretion involves having a written policy that clearly defines what an officer can and cannot do in a particular situation. Confining discretion does not attempt to abolish the use of discretion but only to limit its use to a narrow range of situations by specifying those situations in which they may not use force or conduct a high-speed chase. This approach is consistent with the general view that it is futile and unwise to attempt to abolish discretion completely in any area of criminal justice decision making (e.g., plea bargaining, sentencing) but that rules can effectively control its use.[152] Davis states bluntly that discretion "cannot be eliminated. Any attempt to eliminate it would be ridiculous."[153] Thus, for example, the prevailing standard on the use of deadly force authorizes it only in the defense of life; similarly, high-speed vehicle pursuits of nondangerous offenders are typically prohibited.[154]

Structuring Discretion

Discretion is structured in the Davis model by specifying the factors that an officer should consider in making a decision. As Davis explains, policy should advise officers to "let your discretion be guided by these goals, policies, and principles."[155] High-speed pursuit policies, for example, typically instruct officers to consider road conditions and the potential risk to pedestrians or other vehicles before initiating a pursuit.[156] The Iowa law on domestic violence directs police officers to arrest the person they believe to be the "primary aggressor."[157] These specific guidelines are consistent with the larger goal of striking a balance between the unacceptable alternatives of ignoring discretion altogether or attempting to abolish it. The goal is to limit its exercise as much as possible and to guide it into acceptable application.

Checking Discretion

Discretion is checked in the Davis model by having decisions reviewed by supervisors or even some external authority. Each use of force report is automatically reviewed by higher-ranking supervisors. The knowledge that a review will occur is designed to affect the officer's decision making. Two other mechanisms that are part of the new police accountability further enhance the checking process. Early intervention systems (discussed in Chapter Five) represent a systematic analysis of use of force data to identify patterns that are problematic. Finally, the police auditor form of citizen oversight (discussed in Chapter Six) provides a review by experts who are not themselves sworn officers in the department.

Collateral Contributions

(1) Rules as Statements of Values. Rules have important collateral contributions to good policing. One of the most important is that they embody statements of

values.[158] Use of force policies today typically begin with the statement that the guiding principle is the protection of life. The Kansas City, Missouri, police department policy, for example, declares that "This department recognizes and respects the value and special integrity of human life. In permitting members, with the lawful authority, to use force to protect the public welfare, and for the apprehension and control of suspects, a careful balancing of all human interests is required."[159] Along the same lines, the model policy recommended by the California Peace Officers Association declares, "This department recognizes and respects the value of human life and dignity. Vesting officers with authority to use force to protect the public welfare requires a careful balancing of all human interests."[160]

These statements clarify a department's priorities. The old fleeing felon rule said, in effect, that arrest was the highest priority and that if someone who did not deserve to die is fatally shot, well, that is simply a mistake we have to live with. The defense of life standard reverses the order of priority, saying that the protection of life is paramount and, to that end, we are willing to tolerate the occasional escape of a genuine felon, and in particular we are not going to risk death to an innocent person in the case of a relatively minor offense. Similarly, restrictive pursuit policies communicate the message that the safety of bystanders and other drivers is more important in certain circumstances than the arrest of a fleeing suspect.

The failure of some departments to conform to the new standards is illustrated by the absence of such statements about the priority of preserving life over other police goals. The Philadelphia Police Department, an agency with well-documented accountability problems, did not add a statement about the value of human life to its use of force policy until 1998.[161] The Louisville, Kentucky, police department did not have a clear statement on the value of protecting life as late as 2002.[162]

(2) Rules as Training Tools. Finally, policies serve as important training tools. The standard classroom lecture on human rights, offered in the police academy, suffers from being too far removed from the day-to-day reality of police work on the streets. Preservice lectures, moreover, are easily forgotten once an officer hits the streets. In fact, policing has been notorious for having veteran officers tell the new recruit, "forget all that academy crap, this is how we really do it." The statement of values, by contrast, is embodied in an operational policy that guides an officer in critical incidents and for which the officer knows he or

she will be held accountable. Written policies and their day-to-day enforcement, in short, serve as a very meaningful on-going training for officers.

CONTROLLING POLICE CONDUCT
IN CRITICAL INCIDENTS

Since meaningful controls over the use of deadly force first appeared in the 1970s, the basic administrative rulemaking principle has been extended to other critical incidents involving the use of police powers. The following section discusses some of the most important applications.

The Challenge of Controlling Police Use of Physical Force

Allegations of "police brutality"—meaning the use of excessive physical force—has been as much a volatile civil rights issue as deadly force. The development of meaningful controls over officer use of nonlethal force lagged behind the controls over deadly force and began to reach a comprehensive approach only in the 1990s.[163] In several important respects, physical or non-lethal force is more difficult to control than deadly force. Nonlethal force includes a wide range of behavior and there is considerable ambiguity as to what actions constitute use of "force." With respect to use of deadly force, there is no ambiguity about the fact that an officer fired his or her weapon. Among law enforcement agencies, for example, there is no consensus of opinion on whether a routine handcuffing represents a use of force that should be the subject of an official report. Additionally, the number of nonlethal force incidents in any given year is far higher than the number of shooting incidents, a fact that greatly complicates the task of reporting, reviewing, and controlling such incidents.[164]

Defining Force

A comprehensive use of force policy must first define what actions constitute use of *force*. The Department of Justice report, *Principles for Promoting Police Integrity,* recommends that "agencies should define 'force' broadly, including any and all 'physical efforts to seize, control, or repel a civilian. . . .'"[165] As already mentioned, the use of nonlethal force includes many different actions, each of which has different levels of potential harm to a citizen. For example, is kicking a citizen permissible? Is it permissible to strike a citizen with the knee? Many departments today prohibit strikes to the head because of the potential for serious injury or death, but some others do not explicitly forbid it.[166] Vagueness creates a host of potential problems. One department, for example, approves the use of "leg strikes," but the policy is not clear as to whether this refers to striking the citizen in the leg or using an officer's leg to strike the citizen.[167]

Currently, the most comprehensive use of force policies include any "control of person" action by an officer. The San Diego Police Department policy, for example, requires an officer to file a report after using "Any force option, [or] control hold," in addition to any "weaponless defense technique applied to a person, or any force that causes injury or complaint of injury to either the officer or the subject being restrained."[168] The Miami-Dade Police Department has a similar "control of person" policy, and all reports are entered into the department's early intervention system.[169]

There is no consensus on whether the routine handcuffing of a nonresisting suspect should be defined as force. As a matter of policy, most departments handcuff all persons arrested for a felony. Defining this as force adds a substantial administrative burden because of the number of reports that would result (along with burden of entering these reports into the department's early intervention system).[170] Thus, an effective approach to the control of use of force faces the competing claims of comprehensiveness and efficiency.

As is the case with deadly force policy, many departments have not kept pace with the developing national standards. Investigations by the U.S. Department of Justice under the 1994 "pattern or practice" law have identified specific deficiencies in use of force policies in a number of police departments. Department of Justice (DOJ) investigators found in 2002, for example, that the Detroit Police Department policy "does not define 'use of force' nor adequately address when and in what manner the use of less-than-lethal force is permitted."[171] Similarly, the DOJ found that the Schenectady, New York, use of force policy "contains vague language and undefined terms," it "fails to identify specific uses of physical force that may be prohibited or restricted to limited circumstances," and it does not specify whether officers may use carotid holds, or hog-tying, two types of force that have caused serious injury and death.[172]

Consistency Among All Policies

Developing a coherent use of force policy is complicated by the fact that use of force is typically covered by several different policies, often resulting in a lack of consistency among them. In the Miami Police Department, for example, the DOJ found that whereas one policy embodied a state-of-the-art definition of when force could be used, the policy on arrests contained a vague and far more permissive definition. Additionally, and even more seriously, an Internal Affairs Unit policy called for a review of incidents only in the case of "flagrant use of excessive force." The DOJ concluded that "The MPD . . . fails to provide officers with clear guidance on what constitutes a reasonable use of force."[173]

The inconsistencies in the various Miami use of force policies are, in one sense, a "housekeeping" problem: a failure to review all relevant policies and ensure consistency. In the larger context of achieving accountability, however, housekeeping is not a minor issue. Maintaining a complete and consistent set of policies on all important aspects of police operations is a fundamental management issue. Most departments have yet to abandon the traditional crisis management approach, in which policies are developed in reaction to controversial incidents, and instead take a proactive and comprehensive approach to policy development.[174] The role of police auditors on this critical issue is discussed in detail in Chapter Six, and the larger issue of ongoing proactive housekeeping is discussed in the concluding chapter.

Confining the Use of Force

The second key element in a use of force policy is confining the use of force by specifying the circumstances when it may and may not be used. The prevailing standard is that an officer may use the *minimum amount of force necessary for achieving a lawful purpose.* There is no clear consensus on exactly what these purposes are, however. The Kansas City Police Department reflects the prevailing national standard by specifying that force may be used for four basic purposes: "Members may use department approved non-lethal force techniques and issued equipment to: a. Effect an arrest. b. Protect themselves and others from physical injury. c. Restrain or subdue a resistant individual. d. Bring an unlawful situation safely and effectively under control."[175] Unstated but clearly implicit in this policy is the prohibition on the use of force in response to disrespect to an officer and his or her authority, or what is often called "contempt of cop."[176]

Less-Than-Lethal Weapons

In an effort to reduce the use of deadly force, police departments have adopted various forms of nonlethal force. These include chemical sprays, tasers, and so on. The goal of providing alternatives to deadly force is laudable, but adding nonlethal weapons also creates new policy requirements, as a department must specify the proper use of each nonlethal weapon and provide the necessary training in its use.

The Department of Justice faulted the Detroit police for having only "a limited array of [nonlethal] force options available": a firearm and chemical spray.[177] The Department of Justice found the Buffalo, New York, Police Department deficient with regard to the use of chemical sprays, and directed it to provide eight hours of training on its use, including a "discussion and role plays of situations in which use of CAP [Oleoresin Capsiscum] spray is and is not permissible and how to assess relevant factors before using CAP spray."[178]

The Impact of Less-Than-Lethal Force Policies

In contrast to deadly force policies, there is little research investigating the impact of restrictive nonlethal force policies on the use of force. A 1999 Bureau of Justice Statistics report concluded that "The impact of differences in police organizations, including administrative policies . . . on excessive and illegal force is unknown."[179] That is to say, we have no studies that convincingly demonstrate the impact of a restrictive use of force policy in reducing the use of force and the use of excessive force in particular.[180] The reasons for this lack of empirical evidence are understandable. Deadly force is relatively easy to study because incidents are so few and there is no ambiguity about whether a weapon was discharged and whether someone was shot and killed. Nonlethal force incidents, on the other hand, are numerous and often ambiguous. Also, because low-level uses of force are such a routine aspect of police work, changes in the use of force in a department are likely to be influenced by a variety of factors other than the department's formal policy: the quality of on-the-street supervision, changes in disciplinary practices (with a resulting deterrent effect), and so on.

The lack of solid research on the impact of restrictive use of nonlethal ˌe policies is a serious problem that needs to be addressed by social scienˌts. It is vitally important to confirm the belief that restrictive policies do ˌduce excessive force, and, if so, whether certain policies and procedures are ˌore effective than others.

Reporting and Review Requirements

The current national standard is that officers are required to complete a report after any use of force and that these reports be subject to an automatic review by supervisors. As is the case with defining the use of force itself, this requirement is far more complex than appears at first glance.

One major failing is that some departments require only certain incidents to be reported. The Department of Justice found that in Detroit "officers are not required to report uses of force other than uses of firearms and chemical spray, unless the use of force results in a visible injury or complaint of injury."[181] The 2004 use of force policy in the Las Vegas Police Department requires reports only in cases that involve "death, injury, or complaint of injury," "intentional traffic collision," or discharge of a firearm.[182] In short, the vast majority of force incidents are not required to be reported under these policies.

Additional shortcomings exist with respect to the review of use of force reports. In Schenectady, the DOJ found that although the department's policy required a use of force report for each incident, it did not require supervisors to review or investigate force incidents. Additionally, interviews with both command-level and rank-and-file officers found that, contrary to policy, "officers rarely document uses of force and that supervisors do not enforce the reporting policy."[183]

CULMINATION: THE USE OF FORCE CONTINUUM

The development of use of force policy has reached its culmination in the *use of force continuum*. The continuum is a list of the full range of coercive actions an office can take, from the least to the most serious, with the proper level of force correlated to the action of the citizen.[184] The California Peace Officers Association (CPOA) explains that

A Use of Force Continuum is a visual representation of force options designed to facilitate an understanding of appropriate levels of force by officers. This is accomplished by establishing parameters which exhibit the actions of both the subject and the officer on a comparative scale.[185]

A use of force continuum translates the abstract concept of the minimum amount of force necessary into practical terms that a police officer can readily understand. The CPOA emphasizes that a continuum "should be easily understood and readily recalled by officers under the stress of a confrontation." The Department of Justice recommends a force continuum as a best practice, explaining that "The levels of force that generally should be included in the agency's continuum of force include: verbal commands, use of hands, chemical agents, baton or other impact weapon, canine, less-than-lethal projectiles, and deadly force."[186]

The use of force continuum has been adapted by researchers as a research tool. Professor Geoffrey Alpert developed the analytic framework of the Force Factor, which examines the relationship between the level of force used by an officer and the behavior of the citizen in an encounter.[187] This permits a meaningful analysis of the frequency of the use of excessive force together with a parallel analysis of how often officers use less force than they could have.

Deescalating the Use of Force

An important contribution of the use of force continuum is that it directs officers' attention to the possibility of using lower levels of force. Most continuua list "officer presence" as the least coercive form of force. This advises officers that their physical presence is a form of force that can be used effectively to assert and maintain control over situations in which there is a potential for conflict. As the Kansas City policy explains, "mere police presence often avoids the need for any force."[188]

The Department of Justice agreement with the Buffalo, New York, Police Department directs it to "train all officers in the use of verbal de-escalation techniques as an alternative to the use of CAP spray and other uses of force." The memorandum of understanding with the Washington, DC, police department requires the department to "emphasize the goal of de-escalation and . . . encourage officers to use advisements, warnings, and verbal persuasion when appropriate."[189] Directing officers to deescalate is not only a relatively new idea in

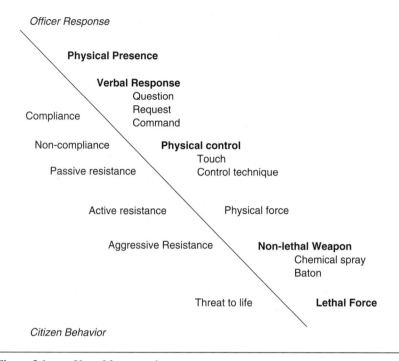

Officer Response

Physical Presence

Verbal Response
Question
Request
Command

Compliance

Non-compliance

Passive resistance

Physical control
Touch
Control technique

Active resistance   Physical force

Aggressive Resistance   **Non-lethal Weapon**
Chemical spray
Baton

Threat to life   **Lethal Force**

Citizen Behavior

Figure 3.1   Use of force continuum.

Source: Adapted from existing department policies.

policing but it addresses a problem that lies at the core of the traditional police subculture. Officers have long regarded challenges to an officer's authority as justification for use of force. This phenomenon is referred to colloquially as "contempt of cop."[190]

A popular technique for deescalating encounters is known as verbal judo, which involves talking to citizen in ways that lead away from a confrontation and possible use of force. One report explains that "Verbal Judo is the principle of Judo itself: using the energy of others to master situations." The goal is to obtain voluntary compliance from people who are hostile or in some way not completely in control of their behavior.[191] Much of verbal judo is simply common sense and a strategy that ordinary people use every day in encounters on the job, among friends, or in the family. Unfortunately, not all people have this kind of common sense, and police officers are no exception. Consequently, both the principle of deescalation and basic tactics have to be taught.

Controlling the Dynamics of Police–Citizen Encounters

Verbal judo and other deescalation tactics rest on the recognition that police–citizen encounters are fluid events that involve a number of different stages and whose outcome is contingent on actions by the parties involved. Thus, it is within the power of the officer to shape the outcome of many (but not necessarily all) encounters. One of the earliest and best discussions of the stages of an encounter is in Peter Scharf and Arnold Binder's book on deadly force, *The Badge and the Bullet*. The four stages are *Anticipation, Entry and Initial Contact, Dialogue and Information Exchange*, and *Final Decision*. Each stage includes actions by the citizen, the perception of those actions by the officer, and the officer's response.[192] There is growing recognition of the potential for training officers to control the development of an encounter and direct it toward a less violent outcome.

The scenario described by Scharf and Binder can be applied to virtually all types of police–citizen encounters. The following section describes the application to encounters with people suffering from mental disorders.

Responding to People With Mental Disorders

Encounters with people with mental disorders are situations policy and training can often assist in avoiding the use of force. In a certain number of encounters, the person displaying erratic or bizarre behavior has a weapon or object that could cause harm; and, in some of these cases, the citizen makes a move that can legitimately be interpreted as a threat to the life of the officer (e.g., swinging a knife) and the officer shoots and kills the person. The postincident review often justifies the officer's action by focusing on the final gesture by the citizen without examining whether the officer could have taken some action earlier in the encounter that would have led it in a different direction.

Currently, the most popular program for improving police response to people with mental disorders is the Crisis Intervention Team (CIT) program developed by the Memphis, Tennessee, Police Department. It involves a collaborative arrangement with mental health professionals and special training for officers in dealing with mentally disturbed persons. CIT has gained national recognition and has been copied by a number of other police departments.[193] A report on Portland, Oregon, explained that "CIT officers receive specialized training in dealing with individuals with mental illness or suicidal ideation, and learn to slow down and deescalate incidents, negotiate with subjects, and respond more flexibly."[194] A number of other departments, including Albuquerque and Seattle, have adopted the CIT program.[195]

CONTROLLING OTHER CRITICAL INCIDENTS

High-Speed Vehicle Pursuits

The first and most important application of the report and review process beyond the basic officer use of force situation involved high-speed vehicle pursuits. As with the use of deadly force, pursuits were essentially uncontrolled until the 1970s. Officers were free to pursue a fleeing vehicle regardless of the circumstances if they simply chose to do so. High-speed pursuits became a part of the culture of policing, with flight defined as a direct challenge to an officer's authority (another version of "contempt of cop"). With the development of media technology (helicopters, more mobile cameras) high-speed pursuits became a part of the popular culture of policing. It is ironic and unfortunate that television has popularized high-speed pursuits just as police departments have begun to limit them.

By the 1970s, an increasing number of experts recognized that vehicle pursuits were extremely dangerous events. The first study to gain national attention, by the Physicians for Automotive Safety, reported the alarming estimates that 20% of all pursuits ended in someone being killed, 50% ended in at least one serious injury, and 70% ended in an accident. Subsequent studies found these estimates to be exaggerated but confirmed the basic point that pursuits are highly dangerous. Alpert and Dunham's study of 952 pursuits in Dade County, Florida, in the mid-1980s found that 33% of all pursuits ended in an accident, and 17% ended with someone being injured (11% ended with an injury to the driver or passenger in the fleeing vehicle, and 2% ended with an injury to the police officer); seven of the 952 ended in a fatality.[196] Although the Alpert and Dunham estimate of accidents, injuries, and deaths was far lower than the original Physicians for Automotive Safety report, it nonetheless confirmed that pursuits are extremely dangerous. Additionally, their study was conducted *after* the Miami-Dade Police Department had instituted a restrictive pursuit policy. Thus, even under these circumstances there was approximately one high-speed pursuit per day.

Unlike deadly force, which has always been a civil rights issue, interest in controlling vehicle pursuits developed as a municipal liability issue. Cities and counties sought to control the costs associated with lawsuits arising from pursuit-related deaths and injuries.

The new controls over high-speed pursuits followed the basic administrative rulemaking model. First, a written policy confines and structures discretion by specifying when and under what circumstances high-speed pursuits are permitted. Current policies typically discourage or explicitly forbid pursuits in situations in which road conditions are dangerous because of rain or snow, or where there is a risk to citizens, such as in school zones or residential neighborhoods. Even more important, many policies forbid pursuits in which the underlying offense is relatively minor, such as a traffic violation. Other high-risk tactics such as ramming the fleeing vehicle or "caravanning" (i.e., pursuits by a long line of police cars) are also prohibited in most policies today.[197]

Officer discretion to pursue is checked by giving supervisors, and in many departments dispatchers as well, explicit authority to order a pursuit terminated. This is an option not available for the control of uses of force for the simple reason that they typically involve split-second decisions, whereas pursuits are events that allow time for the consideration of the circumstances.

Finally, the new pursuit policies require officers to complete detailed reports on each pursuit. These reports are then reviewed by supervisors and in some departments higher command officers.

There is persuasive evidence that controls over pursuit policies effectively reduce the number of pursuits and the number of resulting accidents, injuries, and deaths. Geoffrey P. Alpert found that a new restrictive policy in the Miami-Dade Police Department in 1992 reduced pursuits by 82%; the return to a more permissive pursuit policy in the Omaha, Nebraska, Police Department, meanwhile, resulted in a 600% increase in pursuits. Training also had a dramatic effect on officer attitudes. Prior to training, St. Petersburg, Florida, 58% of officers would pursue in the case of a "low risk" traffic violation; following training, only 24% would pursue in such cases.[198]

Foot Pursuits

Whereas vehicle pursuits have received considerable attention, foot pursuits by officers have been relatively neglected. They typically occur when a police officer stops a motor vehicle and the driver flees on foot. Because they do not involve the use of a vehicle, these incidents are generally not covered by departmental vehicle pursuit policies. The Special Counsel to the Los Angeles

Sheriff's Department found that "In contrast to vehicle pursuits, which are reliably tracked, the LASD does not keep tabs on foot pursuits and currently cannot state how many foot pursuits occur each year, or result in a use of force, or lead to an injury to a deputy."[199]

The Special Counsel's analysis found that foot pursuits are extremely dangerous. About 22% of all LASD shooting cases between 1997 and 2002 (52 of 239 incidents) involved "shots fired by deputies during or at the conclusion of a foot pursuit."[200] Citing several recent cases, the report noted that it was common for officers to charge after a suspect in the dark, losing contact with fellow officers or supervisors, and with no coherent plan of action. Additionally, the LASD was extremely "reluctant" to discipline officers for "tactically reckless foot pursuit that puts the deputy himself in real danger." One lieutenant explained that it was punishment enough for a deputy to later realize that "his ass could have been dead out there," and therefore he would "not act like an idiot again."[201]

For many officers, a suspect fleeing on foot is another form of "contempt of cop," a direct challenge to his or her authority. The LASD Special Counsel's report touched a raw nerve and provoked an extremely hostile reaction from the sheriff's department, suggesting that he had exposed an important aspect of the police subculture.

The Police Assessment Resource Center report on use of force by Portland, Oregon, police officers reached a similar conclusion. It found that the police department's "own training documents" considered foot pursuits to be "one of the most dangerous police actions." The dangers include the minimal reaction time when a suspect stops suddenly and produces a weapon, the danger of being disarmed, the difficulty in communicating with other officers, the risk of fatigue and a physical encounter with the suspect, the problem of a pursuit over difficult terrain, and the risk of an officer not knowing his or her location at the end of a pursuit. Portland's official policy was explicit: "DO NOT ENGAGE IN A FOOT PURSUIT OF AN ARMED SUSPECT. DO NOT PURSUE AN INDIVIDUAL WITH YOUR GUN OUT." In spite of this directive, however, the report found that most of the foot pursuits reviewed involved clear violations of the department's "absolute don'ts."[202]

To control foot pursuits and enhance the safety of both officers and citizens, the LASD Special Counsel recommended the policy of the Collinswood, NJ Police Department, which explicitly prohibits, among other things, pursuits by lone officers into buildings, confined spaces (e.g., fenced-in areas), or wooded areas; pursuits in which the officer loses sight of the suspect and consequently is not sure of his or her whereabouts; pursuits in which the risk to other citizens or other police personnel outweighs the need for immediate apprehension.[203] The Cincinnati Police Department also adopted the Collins-wood foot pursuit policy.[204]

The International Association of Chiefs of Police (IACP) acknowledged the growing awareness of the dangers of foot pursuits in early 2003, issuing a policy paper strongly discouraging them. The IACP advised that "The decision to pursue a fleeing suspect should not be regarded as a required or even a prudent action in all instances." Specifically, it advised that "Unless exigent circumstances, such as an immediate threat to the safety of other officers or civilians, officers should not normally engage in or continue foot pursuits" in which they are "acting alone," going into buildings or other isolated spaces "without sufficient backup," losing communication with other officers or central dispatch, and also in cases in which "the suspect's identity is established where the suspect may be apprehended at a later time with a warrant and there is no immediate threat to the officers or the public."[205]

Use of Police Canines

In the 1990s, police canines emerged as an explosive police–community relations issue. The use of dogs evoked memories of the 1960s civil rights movement, including the famous image of dogs attacking civil rights demonstrators in Birmingham, Alabama, in 1963. Police accountability experts noticed that in many departments there were no formal policies governing deployment of the canine units. The Department of Justice report *Principles for Promoting Police Integrity* unequivocally states that "the use of a canine to attempt to apprehend or seize a civilian is a use of force"[206] and should be incorporated into a department's general use of force policy. Common sense suggests that a dog bite inflicts the same kind of harm as a blow with a police baton.

Efforts to reduce unnecessary harm to citizens from canines have followed the general use of force paradigm: restrictive policies and reporting requirements. In Cincinnati, a new policy mandated by the Department of Justice prohibits canine bites except "where the suspect poses a risk of imminent danger" (e.g., injury), and to call off the dog "at the first possible moment."[207] Additionally, written reports are required of all canine deployments and these reports are to be entered into the risk management system (e.g., early intervention system). The Memorandum of Agreement with the Washington, DC, police department requires similar reforms of canine unit policy.[208] The Los Angeles consent decree requires that all bite incidents, but not mobilizations, be classified as a use of force and entered into the department's early intervention database.[209]

One of the key issues in the deployment of canines is whether the dogs are trained to "find and bark" or "find and bite." In Philadelphia in the 1970s, "Several hundred police canines were trained to bite first and bark second. . . ."[210] The Department of Justice found that the Miami Police Department did not "specify whether it uses a 'find and bite' policy (which allows dogs to bite upon locating a subject) or a 'find and bark' policy (requiring a dog to bark, rather than bite)." Interviews with canine unit officers indicated that in practice the department used a "find and bite" policy. Dogs were trained to bite subjects "regardless of whether the subject is actively resisting or attempting to flee."[211]

A second deployment issue involves when canines can be unleashed. The Memorandum of Agreement between the Department of Justice and the Cincinnati Police Department requires a new policy whereby officers must gain approval from an immediate supervisor before releasing a dog and also must announce "loud and clear" to a suspect that a canine deployment is imminent.[212]

A new restrictive canine unit policy resulted in a sharp decline in deployments and bites in the Los Angeles Sheriff's Department. One key change was a limit on deployment of canines on auto theft suspects. Data indicated that many of these cases involve teenage joy rides that do not pose serious dangers to either officers or the public. The ban eliminated about 25% of all canine deployments.[213] Data on the impact of the new LASD canine policy is in Chapter Six (pp. 150-151), in the context of a discussion of the role of the Special Counsel as a police auditor.

The Display of Weapons

The display of an officer's firearm, while not technically a use of deadly force, is both a risky action because of a possible accidental discharge and an expression of police powers that is highly offensive to citizens, especially African Americans. It is a blatant reminder that the officer possesses the ultimate power of life and death. In Cincinnati, this practice was apparently fairly common and was a major grievance in the African American community prior to the April 2001 riots.[214]

The first expression of concern that the display of weapons by police officers should be limited appeared in the 1977 Police Foundation report on deadly force.[215] The Memorandum of Agreement between the Department of Justice and the Washington, DC, police prohibits officers from "unholstering, drawing, or exhibiting a firearm unless the officer reasonably believes that a situation may escalate to the point where deadly force would be authorized." Additionally, officers must "complete a Use of Force Incident Report immediately following the drawing of and pointing of a firearm at, or in the direction of, another person."[216] Similar policies have been developed in Cincinnati and the Miami Police Department as a result of Department of Justice intervention.

INVESTIGATING USE OF FORCE AND OTHER CRITICAL INCIDENTS

The review of use of force reports is the second part of the modern use of force policy paradigm. Although it might seem that investigating a use of force incident is a fairly simple and straightforward process, in fact it is extremely complex and problematic. The following section examines some of the most important issues related to the review of reports and the investigation of use of force incidents.

Centralized Versus Decentralized Investigations

The first issue regarding the review of force reports is where primary responsibility for the review should lie. There is a general consensus of opinion that, while an officer's immediate supervisor has important responsibilities on this issue, he or she should not be the sole reviewer. To avoid possible favoritism because of friendship, other higher-ranking officials should review incident reports. The most important question is whether primary responsibility should be centralized in the Internal Affairs or equivalent unit or decentralized and handled at the precinct level. There is no consensus on this question at present. Centralization has its obvious merits. It ensures a consistent enforcement of department policy by commanders who are not likely to be influenced by personal relations with the officers in question. It also means that the chief and other top commanders have a complete picture of critical incidents throughout the agency. At the same time, however, decentralization has important advantages.

Captains, lieutenants, or even sergeants at the precinct level can respond more quickly to particular cases. The amount of paperwork and resulting delays can be greatly reduced. Giving lower-level supervisors responsibility for reviewing force reports, moreover, reinforces the idea that they have a major role in the larger accountability program in the department.

Some departments have resolved the issue by dividing responsibility. Serious use of force incidents are automatically handled by a centralized unit, whereas precinct-level commanders retain responsibility for less serious cases, including even imposing discipline on officers guilty of misconduct. The consent decrees negotiated by the Department of Justice in recent years require centralization. In large part this is because the departments being sued have had such dismal records with regard to accountability. In Los Angeles, for example, force investigations are now centralized in the Operations Headquarters Bureau.[217]

Immediate "Roll Outs"

To ensure the integrity of force incident reports, immediate "roll outs" to serious use of force incidents are increasingly used, involving either an immediate supervisor, an internal affairs unit officer, or even someone from an external agency. This practice is designed to ensure that officers at the scene do not alter the physical evidence or conspire to create a cover story.

The Cincinnati Memorandum of Agreement requires officers to notify their supervisors after any use of force, and that supervisors promptly respond to the scene. Additionally, the Internal Investigations Section is required to respond to the scene of "serious" force incidents and all canine bites that cause injury or require hospitalization.[218] In the Los Angeles Sheriff's Department (LASD) the Office of Independent Review, a team of seven attorneys, rolls out to all shooting incidents.[219] In Miami, Florida, the State Attorney's Office, which has the authority to prosecute an officer for homicide, responds to all shooting incidents in which a citizen or officer is fatally shot or wounded.[220]

Several other aspects of the immediate aftermath of a shooting incident are also subject to new rules. In the LAPD, the consent decree requires that all officers and witnesses are to be "separated immediately" after a shooting incident.[221] This is designed to prevent officers from colluding to create a common version of the incident that justifies the shooting. The decree also directs the department to negotiate with the police union to secure a requirement that, in the case of shootings involving more than one officer, each officer be represented by a different attorney.[222] Some departments conduct formal debriefings of officers involved in use of force incidents. The settlement agreement with the Riverside, California, Police Department, for example, requires a debriefing after each "critical incident," defined as any unplanned event that threatens community peace and safety.[223] Finally, higher-ranking officers who respond to force incidents are required to evaluate whether or not an immediate supervisor was present.[224] Investigation of the role of supervisors in critical incidents is discussed in more detail below.

Ensuring Unbiased Investigations

Ensuring fair and unbiased investigation of use of force incidents is another critical issue that is increasingly addressed through specific policies and procedures. The belief that investigations are biased and essentially "whitewashes" of official misconduct has been a major issue for civil rights activists since the 1960s. The 1992 Kolts report on the Los Angeles Sheriff's Department found "explicit and implicit biases against civilian complainants at every level of the complaint process." These problems included investigations being conducted by the supervisor of the officer under investigation, with resulting evidence of bias, investigations being "closed before completion—at times under highly suspicious circumstances," and complaints that are "corroborated by physical evidence and independent witnesses are frequently not sustained."[225] The first report by the court-appointed monitor in Philadelphia found that investigators failed to follow leads and take obvious investigative steps, and had an ingrained tendency "to view the case only from the officer's perspective."[226]

As a corrective measure, investigators in the LAPD are now directed not to ask "leading questions" of either officers or citizens, nor to give an automatic preference for the statements of officers over those of citizens.[227] Other new policies and procedures designed to ensure fair investigations are discussed in Chapter Four with respect to citizen complaints.

Witness Officers and the Code of Silence

One of the greatest obstacles to the investigation of misconduct incidents and to police accountability generally has been the refusal of involved officers to give honest answers to investigators. Officers who witness events

---

2.11.2 RESPONSIBILITIES OF ALL MEMBERS

Members are required to immediately notify their commanding officer or civilian supervisor of violations of orders, policies or procedures, disobedience of orders by other members, or mismanagement related to the effective and efficient operations of the Department. The supervisor or commanding officer must document specific violations. Members inhibited by the chain of command from reporting misconduct are required to submit the information directly to the Chief of Police or to the Commander of the Professional Standards Unit in writing. Members are prohibited from taking punitive action or discriminating against any member who reports a violation under this policy.

Figure 3.2   Louisville Police Department SOP: "Discipline" 11/16/03

under investigation often refuse to either report what they observed or give complete and honest answers to investigators. This problem is generally referred to as the "code of silence," and it has been identified in innumerable reports as an impediment to the investigation of officer misconduct.[228]

Some new policies and procedures have been developed that address this long-standing problem. An increasing number of departments have a policy explicitly directing officers to report and testify accurately about misconduct by other officers. Figure 3.2 contains the policy recently adopted by the Louisville, Kentucky, Police Department.

Some attention has also been given to providing protection for whistle-blowers, officers who voluntarily come forward and report misconduct by other officers. The consent decree with the Los Angeles Police Department directs the Inspector General for the LAPD to receive anonymous complaints from officers and not be compelled to disclose their names.[229] There are few other programs specifically designed to protect whistleblowers, however. In April 2004, Rutgers Camden Law School and the ACLU cosponsored the first conference to address this issue.[230]

A Neglected Issue: The Police Officer's Bill of Rights

Another problem that has not received sufficient attention involves potential obstacles to investigating misconduct posed by legal protections of the rights of police officers. About 14 states have laws known as the Law

Enforcement Officers Bill of Rights (or some variation of that), that provide specific due process protections for officers under investigation. Additionally, an unknown number of local collective bargaining agreements provide similar and often more extensive protections. These protections typically include the right to notice of the charges; the right to an attorney; a requirement that interviews be conducted at a reasonable time and place; prohibitions on threats, coercion, or retaliation; and so on.[231] Some local collective bargaining agreements contain provisions protecting the rights of officers under investigation that provide far more protective of officers than do the state statutes.[232]

An analysis of the 14 state statutes found that most of the provisions are legitimate due process protections that pose few obstacles to investigations. However, a few provisions are pernicious. Some union contracts include waiting periods that prevent investigators from interviewing an officer for up to 48 hours. Two state laws, meanwhile, prohibit interviews of officers by nonsworn officials, a provision that precludes investigations by a citizen oversight agency. More research is needed on the nature of police officers' bills of rights and their impact on day-to-day investigations of misconduct.

The Framework for Investigations

A critical but neglected aspect of the investigation of force incidents is the framework that guides investigations. A report on use of force in the Portland, Oregon, Police Bureau by the Police Assessment Resource Center (PARC) points out that traditionally shooting incidents are investigated as homicides. This framework immediately focuses the investigation on the narrow (albeit important) question of whether or not criminal charges should be filed against the officer who did the shooting. The PARC report forcefully argues that this approach is no longer "consistent with best practice," and points out that "numerous agencies" have abandoned the practice.[233] PARC argues that "while Homicide investigators are typically well qualified to conduct a *criminal* investigation, they lack either the training or perspective necessary to investigate officer-involved shooting or in-custody death cases from an *administrative* and *tactical* point of view."[234] Investigations should address "the policy and training aspects of such cases."[235] A broader framework can "use the incident as a learning tool . . . to inform and improve the department's policies, procedures, training, and management." Current Portland policy requires that after action reports include a narrative of the incident; a conclusion about whether the officer's action was in compliance with department policy; a critique of whether the incident was handled well; and any appropriate recommendations regarding possible changes in departmental policy, procedure, or training.[236]

The PARC report reflects the recommendation of the Department of Justice that "To the extent possible, the review of use of force incident and use of force reports should include an examination of the police tactics and precipitating events . . . so that agencies can evaluate whether any revisions to training or practices are necessary."[237]

A broader framework for investigations can also inquire into the conduct of an officer's immediate supervisor in a force incident. It is likely that many questionable incidents could be avoided if the supervisor had acted properly, either in the incident itself or in terms of general supervision prior to the incident. The LAPD consent decree addresses this problem by requiring that when reviewing critical incidents, commanders "shall analyze the circumstances surrounding the presence or absence of a supervisor." Additionally, "such supervisory conduct shall be taken into account in each supervisor's annual personnel performance evaluation."[238]

Consistency in Discipline

In the end, officers learn that a department is serious about accountability when they see critical incidents investigated thoroughly and discipline actually imposed for violations of policy. Nothing undermines a use of force policy more quickly than the failure to discipline an officer who clearly used excessive force or who violated some departmental policy. The Philadelphia Integrity and Accountability Office found many instances of officers not disciplined even though the department had sustained the allegations against them.

A related problem is a lack of consistency in discipline. One of the greatest causes of morale problems among rank-and-file officers is the perception that some favored officers escape proper discipline. To correct this problem some departments have adopted a discipline matrix, a schedule of discipline similar to sentencing guidelines in criminal courts that prescribes a disciplinary action based on the seriousness of the immediate incident and an officer's disciplinary record.[239] A few departments have developed discipline matrices but there have been no studies of their impact or the best form they should take.

THE CHALLENGE OF IMPLEMENTING
USE OF FORCE POLICIES

The formal elements of a comprehensive use of force and critical incident reporting system are clear: written policies clearly indicating approved and unapproved behavior, required reports after each incident, and automatic review of all incident reports by supervisors. What is less clear is how to implement a comprehensive policy in a department in which one does not exist and in which the prevailing organizational culture does not embody a commitment to accountability. Changing the organizational culture of policing, in fact, may be the greatest challenge and impediment to the new police accountability.

The Struggle for Change: The Case of Philadelphia

The Philadelphia Police Department represents an excellent case study in the difficulties in changing the organizational culture of a department. The problems are well documented because of the reports by the Integrity and Accountability Office (IAO). The IAO was created as part of the settlement agreement ending a suit brought by the ACLU, the NAACP, and the Barrio Project against the Philadelphia Police Department, and it functions as a form of the auditor model of citizen oversight of the police (see Chapter Six).[240] The IAO Director has the authority to investigate accountability related issues such as the use of force and disciplinary practices and to issue public reports. The IAO reports to date are particularly valuable in not only documenting the problems in the Philadelphia Police Department, but in illustrating how accountability issues are embedded in deeply ingrained administrative practices. These practices, in turn, reflect both formal policies and informal practices that, together, constitute a distinct organizational culture.

The 1999 IAO report on use of force in the Philadelphia Police Department provides a revealing picture of the department's norms. Ellen Ceisler, director of the IAO, found that as recently as the 1970s, "The police culture at that time was completely intolerant of internal reporting of excessive force." Veteran officers "recall this era as the 'wild west,' 'open season' and a 'free for all'" with respect to the use of force. Some officers "recalled the use of plant guns and other weapons as an accepted, albeit unofficial practice." The canine unit was shaped by the same norms. In the 1970s, "Several hundred police canines were trained to bite first and bark second. . . ."[241]

Although the Philadelphia Police Department has in place many of the formal requirements of a use of force reporting system, the IAO found a deeply ingrained resistance to implementing the system. Some commanders, for example, regard controls over use of force "with resentment, cynicism and suspicion, viewing them as burdensome and unnecessary chores. . . ." Moreover, the IAO report continued, "A number of supervisors and commanders we interviewed did not seem to understand the goal or purpose of the use of force reporting and investigation policies and procedures, viewing the process as a waste of time."[242] Many commanders viewed the reporting requirements and resulting investigations as "inherently punitive." Supervisors are reluctant to "jam up" (i.e., investigate) officers "who they feel were just doing their jobs."[243]

The IAO investigation found a number of deficiencies in use of force reporting practices. Many reports suffered from a "Lack of detail regarding the nature and extent of the subject's injuries." A report, for example, might mention a "head injury," without specifying whether that involved a scratch or a broken skull. The details of force incidents were "routinely sparse, vague, inaccurate, or incomprehensible." Even more fundamental, "The type of force [used] was not always disclosed" in official reports, and the names of officers involved not always provided.[244]

The IAO also found that low-level uses of force (grabs, pushes, shoves) were required to be reported but in practice were not subject to systematic review by supervisors. In fact, the department had no formal policy "regarding supervisor/commander obligations to review use of force notifications and incidents." The Internal Affairs Bureau (IAB) had its own internal guidelines for investigating use of force incidents, but they were neither formal written policies nor official department policy.[245]

Astonishingly, the official Discipline Code of the Philadelphia Police Department "does not include a provision which specifically addresses inappropriate use of force." Consequently, disciplinary actions are usually brought as charges of conduct unbecoming an officer or neglect of duty. Even then, many inappropriate uses of force incidents were never brought forward for disciplinary action. The IAO office found "Numerous cases" of allegations of physical abuse that were sustained by IAB but then not included in the formal charges prepared by the commander who prepared the case for review by the Police Board of Inquiry (PBI).[246]

The IAO concluded that, with respect to discipline, the Philadelphia Police Department remained deeply resistant to change. "With few exceptions," the formal Disciplinary Code and the informal disciplinary process "has remained fundamentally the same for decades."[247] Despite this gloomy assessment, the IAO did find some signs of progress. The 1996 consent decree had some positive effects on use of force reporting practices, including examples of "innovative and productive uses of use of force information" by the department. For example, under a new Case Review Program, three members of the IAB conduct review sessions with officers whose records indicate a pattern of force incidents.[248]

Philadelphia may be an extreme case in terms of the apparent resistance to the basic principle of seriously investigating critical incident reports, but the problem of changing established habits exists in all police departments—in all large organizations, for that matter. With this in mind, we need to conclude the discussion of use of force reporting with the recognition that, while we now know what needs to be done to reduce misconduct, we face a major challenge in terms of how to implement the necessary reforms.

CONCLUSION

The police have awesome powers unrivaled by any other public officials: to deprive people of their liberty, to use physical force against resisting clients, and ultimately to take human life. Ensuring that these powers are used only when absolutely necessary and without bias against any group is a matter of the highest priority. After many decades of shameful neglect, the police have developed a process for controlling police use of force. The essential features of that process are simple: specifying when force can be used, requiring officers to complete a report on each force incident, and reviewing each report. As this chapter has explained, however, each of those elements is extremely complex and filled with problematic issues. Only in recent years have police departments begun to address all of the relevant issues, often under the compulsion of the U.S. Department of Justice and the federal courts. Much remains to be done. The primary issue confronting the police today is one of organizational change: how to implement the necessary changes in a complex bureaucracy and ensure that they become a part of the organization's operational life. The problem of effecting organizational change reappears in the two chapters that follow.

⋈ FOUR ⋊

AN OPEN AND ACCESSIBLE
CITIZEN COMPLAINT SYSTEM

————— ◆ —————

CITIZEN COMPLAINTS AND
THE NEW POLICE ACCOUNTABILITY

The citizen complaint process in the Oakland Police Department was almost completely dysfunctional. The court-appointed monitor's February 2004 report found systemic failures: Formal deadlines for completing phases of complaint investigations were "sparse"; even existing deadlines were frequently "not met," with investigations "delayed or halted for reasons not related to complexity of the case"; even more seriously, officers were often not disciplined for complaints that were in fact sustained; some cases were "filed" while civil suits were pending but never reopened. The monitor concluded that "The most striking overall finding is the failure of OPD's structure as a whole to support the internal investigations process."[249]

The problems in the Oakland citizen complaint process are typical of similar problems that historically have plagued American police departments and been the source of serious police–community relations tensions. An open, accessible, and accountable citizen complaint process is a key component of the new police accountability. A citizen complaint process is a mechanism by which a police department can make itself accountable to the people it serves: by hearing their complaints, investigating them, and, where appropriate, disciplining officers guilty of misconduct. Even when complaints are not sustained, citizens want an opportunity to be heard and treated in a respectful manner.

Along with use of force reports, citizen complaints are a valuable indicator of officer performance and are incorporated into all early intervention systems (see Chapter Five) that seek to identify patterns of officer performance problems. To this end, the Department of Justice report, *Principles for Promoting Police Accountability,* recommends that police departments "provide a readily accessible process in which community and agency members can have confidence that complaints against agency actions and procedures will be given prompt and fair attention."[250]

The operating principles of an effective citizen complaint procedure are openness, integrity, and accountability. *Openness* means that the process makes an effort to inform citizens about the complaint process and to receive all citizen complaints, no matter how frivolous some might seem. *Integrity* means that the complaint investigations are conducted in a manner that is thorough and unbiased. *Accountability* means that the complaint process itself is subject to review to ensure that it operates properly and effectively.

Historic Problems With Complaint Procedures

An open, accessible, and accountable citizen complaint process represents a fundamental shift in the way police departments respond to citizen complaints. Historically, the police have been extremely hostile to complaints, denying that the alleged misconduct occurred, often rebuffing citizens attempting to file complaints, failing to investigate complaints in a thorough and fair manner, and not disciplining officers who are in fact guilty of misconduct. This defensive posture has been one part of the pervasive refusal to take an aggressive approach to reducing officer misconduct. Numerous reports over the years have documented the failures of how police internal affairs units have handled citizen complaints. In the 1960s, both the President's Crime Commission and the Kerner Commission found police departments that either had no formal citizen complaint process or turned away citizens attempting to file complaints.[251] In the 1970s and 1980s, various reports by the ACLU, the U.S. Civil Rights Commission, and other organizations found evidence of unacceptable complaint review procedures.[252] As recently as 1992, the Kolts report on the Los Angeles Sheriff's Department found "explicit and implicit biases against civilian complainants at every level of the complaint process." The problems included investigations being conducted by the supervisor of the officer in question, with resulting evidence of bias, investigations being "closed before completion—at times under highly suspicious circumstances," and a failure to sustain complaints that are "corroborated by physical evidence and independent witnesses."[253]

In 2003, the court-appointed monitor overseeing the Los Angeles Police Department found that people seeking to file complaints were turned away, subjected to long delays, and even harassed by LAPD officers. LAPD officers failed to follow mandated procedures for handling citizens wishing to file complaints in 11 out of 19 cases (for a failure rate of 57%). The court-appointed monitor found this failure rate "shocking," "outrageous," and "discouraging."[254] In one case, an undercover police officer posed as a juvenile complaining about an officer's misconduct. The desk sergeant took an extremely long time handling the complaint, extended it beyond 10:00 p.m., and then detained the complainant for a curfew violation. The failure of the LAPD to provide a receptive complaint process is especially shocking in light of the intense public scrutiny of the department as a result of the consent decree and—presumably—officers' knowledge that compliance with the decree would be monitored by one device or another.

Because of the historic problems with internal police complaint procedures, civil rights activists have long demanded the creation of external citizen complaint procedures, usually in the form of a civilian review board. Their assumption has been that an external agency will conduct more thorough and fair investigations because it is independent of the police department and the police officer subculture. This assumption has been only partially fulfilled, however. Although external complaint agencies now cover virtually all big city police departments in the United States, only a few have clearly demonstrated that they in fact do a better job of handling complaints than police departments. Chapter Five discusses police auditors, a new form of external citizen oversight that has shown considerable promise as an effective alternative to the traditional civilian review board.[255]

The Lack of Professional Standards

Establishing a citizen complaint procedure that meets the standards of openness, integrity, and accountability is far more difficult than most people imagine. With respect to an external complaint procedure, for example, it is not sufficient merely to have an agency that is nominally independent of the police department it serves. An effective complaint procedure, whether external or internal, depends on a host of administrative details relating to how complaints are received, recorded, and classified; how complaints are investigated; and finally, whether the process itself is accountable to the public.

At present there are no recognized professional standards for complaint procedures. Neither the law enforcement profession nor the new citizen oversight professional community have developed a set of professional standards for complaint procedures. One cannot find, for example, a recommended standard on such a basic issue as the appropriate number of complaint investigators for a police department of a given size. The current accreditation standards promulgated by the Commission on Accreditation for Law Enforcement Agencies (CALEA) specify that departments should have a formal complaint process, but they provide absolutely no details on such critical questions as minimum staffing levels.[256] The International Association of Chiefs of Police (IACP) policy paper on *The Investigation of Misconduct* addresses a number of legal issues surrounding complaint investigations but ignores most of the administrative issues related to a complaint process.[257]

The only existing set of standards is the proposed Model Citizen Complaint Procedure developed by this author and published in *Police Accountability: The Role of Citizen Oversight.*[258] That model, together with additional new issues, forms the basis for most of the discussion in this chapter.

The void left by the absence of a set of professional standards has been partially filled in two ways. First, a number of citizen oversight agencies have developed detailed policies and procedures of their own.[259] Second, the consent decrees negotiated by the Justice Department settling pattern or practice suits mandate a number of specific policies and procedures for handling citizen complaints. In this regard, law professor Debra Livingston points out that one of the positive benefits of federal pattern or practice litigation may be to stimulate "the articulations and dissemination of national standards governing core police managerial responsibilities."[260]

The New Paradigm

On a more optimistic note, an increasing number of police departments have developed more open and accessible citizen complaint procedures, particularly through their Web sites. The Web site of the Portland, Oregon, Police Department, for example, has a button prominently displayed on the upper left-hand corner reading, "I want to . . . file a complaint/commendation."

Clicking on the button leads to the complaint form and a wealth of information about the complaint process. The Washington, DC, police department Web site explicitly states that people who are unhappy with the way they were treated "are encouraged to file a formal complaint." And the Springfield, Missouri, police Web site declares, "Your complaint is IMPORTANT."[261]

The new paradigm for understanding citizen complaints is that they are valuable *management information:* indicators of officer performance that the command staff needs to know about. The old view was that they are hostile attacks on the department that should be resisted at all costs. To be sure, many complaints are without merit, but departments increasingly recognize that they need to be open and receptive to all complaints, and then sort out the valid ones later. Complaint data are entered into a department's early intervention system (see Chapter Five), where they can be correlated with other performance indicators to identify officers with performance problems that need to be addressed.

THE CITIZEN COMPLAINT PROCESS

The Oakland monitor's comment about the failure of the police department's "structure as a whole to support the internal investigations process" highlights the crucial point that a complaint review procedure is a management *process* that includes many different stages, each of which involves difficult issues that need to be addressed by formal standards. This section examines the most important aspects of a citizen complaint procedure. It discusses particular stages of the process and describes the emerging best practices standards. The aspects of the complaint process fall into three broad categories: (a) community outreach to inform the public about the complaint review system, (b) the complaint investigation process, and (c) staffing and managing the complaint review system.

Community Outreach

Publicizing the Complaint Process

The starting point for an open and accessible complaint procedure involves a sincere effort to publicize the process and inform citizens about how to file a complaint. A complaint procedure is an important part of the organizational "face" that a department presents to the public. The information should include a description of the formal complaint process, how and where to file a complaint, and what a complainant can expect in the way of possible outcomes. Complaint forms should be readily available at convenient locations throughout the community, electronically, and in the appropriate languages (see below). Historically, departments have provided as little information as possible. An evaluation of the Albuquerque Police Department (coauthored by the author of this book) found that although the department had printed brochures in both English and Spanish explaining the complaint process, they were not distributed throughout the community and were in fact unknown to community leaders.[262]

The settlements negotiated by the U.S. Department of Justice all require improvements in how police departments publicize the complaint process and receive complaints. The consent decree covering the New Jersey State Police requires that the department "develop a program of community outreach to inform the public about State Police functions and procedures, including motor vehicle stops, searches and seizures, and the methods for reporting civilian complaints or compliments regarding officers." It requires that complaint forms and informational materials be available "at State Police headquarters, all State Police stations, and such other locations around New Jersey as it may determine from time to time," and that information be provided on the Internet, and at state-operated rest stops located on limited access highways.[263] In addition, the New Jersey State Police must place posters around the state indicating the availability of the 800-number "Hotline" for filing complaints.[264] Along the same lines, the consent decree with the Cincinnati Police Department requires the city to "make complaint forms and informational materials available at City Hall, CCA [the Citizen Complaint Authority], all CPD [Cincinnati Police Department] district stations, libraries, the internet, and, upon request, to community groups and community centers." In addition, at each police district station, the police department is required to permanently post a placard describing the complaint process and include the relevant phone numbers.[265]

Reaching People With Limited English Proficiency

As immigration continues to transform the population of the United States, virtually all communities contain recent immigrants who either have very limited command of the English language or do not speak English at all. Many do not understand the nature of the complaint process, thinking that it is similar to the criminal process requiring an attorney. Finally, many are extremely fearful of police retaliation because of experiences in their home countries. To serve these communities, citizen complaint procedures need to provide informational material in all the languages spoken in the local community. The Web page of the Washington, DC, Office of Citizen Complaint Review (OCCR), for example, has a prominently displayed button marked "En Espanol," which leads to pages in Spanish explaining the complaint process. The OCCR explains that "To make OCCR's services accessible to non-English speaking members of the District's diverse population, OCCR has prepared complaint forms and informational materials in 13 languages other than English, including Chinese Mandarin, French, Haitian Creole, Japanese, Russian, and Vietnamese."[266]

The Responsibilities of Police Officers

Rank-and-file police officers have an important role with regard to informing citizens about the complaint process. Traditionally, officers on the street have responded in an unprofessional and hostile manner when citizens express a desire to file a complaint. They have provided no information about the complaint process, told them it would do no good to file a complaint, or possibly even threatened the citizen with some kind of retaliation. In the 1960s, the refusal of officers to even provide their name and badge number to citizens was a major source of police–community relations tensions.[267]

The new police accountability requires officers to respond in a polite, professional, and informative manner to all citizens, regardless of the circumstances of the encounter, and to respond professionally when citizens say they want to file a complaint. The Cincinnati Memorandum of Agreement, for example, requires "all officers to carry informational brochures and complaint forms in their vehicles at all times while on duty. If a citizen objects to an officer's conduct, that officer will inform the citizen of his or her right to make a complaint. Officers will not discourage any person from making a complaint."[268] Along the same lines, New Jersey state troopers are now required "to carry fact sheets and complaint forms in their vehicles at all times while on duty." In addition, the consent decree specifies that "The State Police shall require all troopers to inform civilians who object to a trooper's conduct that civilians have a right to make a complaint." Finally, "The State Police shall prohibit state troopers from discouraging any civilian from making a complaint."[269]

Multiple and Convenient Methods of Filing Complaints

An open and accessible complaint process includes different ways of conveniently filing a complaint. Traditionally, complaints could only be filed at police headquarters, and because headquarters is a hostile and threatening environment to many complainants in particular, this discouraged some complainants.

In recent years, police departments have expanded the options for filing complaints. Some departments have moved their internal affairs unit to a separate location. Others have made it possible to file complaints at the mayor's office or other official government facility. Some accept mail-in or phoned-in complaints, and an increasing number now accept electronically filed complaints. The Los Angeles Police Department and the New Jersey State Police maintain toll-free 800 telephone numbers to receive complaints. The LAPD consent decree requires that complaints can be filed in person, by mail, telephone, fax, or electronic mail.[270] In person complaints can be filed at LAPD headquarters, any department station or substation, or the offices of the Police Commission or the Inspector General.[271]

The Issue of Anonymous Complaints

Traditionally, the police did not accept anonymous citizen complaints. Many police departments and some citizen oversight agencies, in fact, require the complainant to sign the complaint form. This practice has begun to change. Some police departments and oversight agencies have begun accepting anonymous complaints, whereas others are compelled to do so by consent decrees (e.g., the New Jersey State Police, Los Angeles).[272]

Whether or not to accept anonymous complaints depends on how one views the complaint process. If it is narrowly defined as analogous to the criminal process, with the goal of adjudicating guilt or innocence, then it follows that complaints should be signed. But if complaints are viewed as management information that helps a department proactively address potential performance problems, then it makes sense to accept anonymous complaints. The latter view has emerged as the new standard and has been incorporated into most of the settlements negotiated by the U.S. Department of Justice.

Citizen Inquiries, Questions, and Complaints

People contact citizen complaint offices with all sorts of inquiries, questions, complaints, and issues. Many of these do not involve an actual complaint about an officer's conduct. Some do not even involve the police department. Only some of these contacts become a formal complaint that needs to be investigated. Many are requests for information. Others involve misunderstandings about the law or police procedure: People arrive at a police station to complain about an arrest or citation not understanding that what they did is against the law; others complain about being handcuffed, not understanding that it is department policy to handcuff all persons arrested for a felony. Some people are distraught and simply want to vent their anger or frustration at some public official. (In fact, some police officials have quietly begun to appreciate having an external oversight agency because, as one put it, the police department no longer has to handle all the "crap.") In 1998, the now-abolished Minneapolis Civilian Review Authority (CRA) had more than 800 contacts with citizens, but only 136 of these became formal complaints that were investigated.[273] In 2003, meanwhile, the Washington, DC, OCCR had a total of 613 contacts with citizens, but only 361 became formal complaints.[274]

Most complaint review procedures have not taken inquiries seriously and have not bothered to record them. This approach represents an inappropriately narrow definition of a complaint procedure's role. Handling all citizen inquiries, no matter how irrelevant or ludicrous they may be, is an important public service. Each inquiry represents an unhappy citizen who deserves an explanation or just a chance to express his or her frustration. The Boise Ombudsman recognizes the importance of such inquiries, and its official procedures state that it "will make every attempt to answer or resolve Citizen Inquiries."[275]

The Complaint Investigation Process

For decades, the main criticism of police internal affairs units voiced by civil rights groups is that they do not investigate complaints in a thorough and unbiased fashion. External citizen review of complaints is their strategy for ensuring better investigations. As already mentioned, however, until recently little attention has been paid to developing standards that would ensure thorough and fair investigations, by internal or external review agencies. The following section examines the specific issues—and the necessary standards—related to thoroughness and fairness.

The IACP concept paper on *Investigating Employee Misconduct* is typical of the failure of the law enforcement profession on these issues. The paper gives considerable attention to a set of essentially legal questions related to the requirements for terminating an employee and the use of polygraphs or other tests, but says virtually nothing about the administrative procedures necessary to ensure thorough and fair investigations.[276]

Accepting, Screening, and Classifying Complaints

Police departments have traditionally kept the official number of complaints low by either not accepting complaints at all, not officially recording them, or deliberately misclassifying them. In some instances, complainants have been actively turned away when they walk into a police station, deliberately provided misinformation about where complaints can be filed, or even threatened with arrest. In one of the more notable cases, Rodney King's brother was threatened with arrest when he went to file a complaint at an LAPD precinct station. NBC sent an investigator into New York precinct stations with a hidden video camera to ask for a copy of the complaint form. In about half the precincts, he was not given the form and in several, treated very rudely.[277]

The issue of not accepting some complaints is complicated by the fact that many are indeed patently frivolous. Many citizen oversight agency staff interviewed by this author tell stories about repeat clients who file numerous complaints, almost all without merit.[278] No complaint procedure can effectively investigate every complaint it receives. It is necessary to have a process for screening complaints, quickly weeding out those that are clearly frivolous and concentrating on the serious ones.

The solution to this problem is to accept and record all complaints without judging their merits and then screen them on the basis of clearly articulated criteria. Accepting all complaints enhances accountability by ensuring that a department is not engaging in a practice of turning away complainants. Subsequent screening, meanwhile, facilitates efficiency by allowing the agency to concentrate on the meritorious and serious complaints. In addition to dismissing frivolous complaints, the screening process should also involve separating the more serious complaints from the less serious. This allows an agency to concentrate its resources on full investigations of the serious complaints and handle the less serious ones in a more expeditious manner. The Boise Ombudsman allows Class II complaints (which include allegations of general demeanor without the use of force and selective enforcement)[279] to be referred back to the police department for investigation by a supervisor. The complainant, however, must agree to this referral. This requirement helps to ensure that the complaint process serves the wishes of complainants and does not improperly shunt allegations of serious misconduct into a lower status.

A related problem is that some serious complaints will be improperly classified and placed in a lower category. In one of its first important actions in 1993, the San Jose Independent Police Auditor (IPA) found that the San Jose Police Department was improperly classifying some serious complaints. As a result of the IPA's recommendation, the number of Class I complaints rose significantly the following year (and then declined in 1995 following another change in classification procedures designed to streamline the process).[280]

The Issue of Withdrawn Complaints

In many cases, citizen complaints are dismissed after being initially accepted because the complainant withdraws, refuses to cooperate further, or cannot be located (often because he or she fails to return phone calls). The New York City Civilian Complaint Review Board (CCRB), for example, administratively closes more than half of all the initial complaints it receives each year, "truncating" 51.6% between January 2002 and mid-2003, for example.[281]

Complainant withdrawal is an extremely difficult issue because it can be the result of any one of several factors, one of which involves the quality of the complaint process. Some complainants undoubtedly decide that it is not worth their time and effort. Others may realize that they were at least partly culpable for whatever happened and conclude that they will not prevail. In New York City, complainants withdrew in 12.8% of all complaints, were unavailable in another 11.9%, and were uncooperative in 27%.[282]

At the same time, some complainants may withdraw because complaint investigators are unresponsive or even hostile. It is very easy for an investigator to communicate an unsympathetic message to a complainant, either subtly or crudely. In addition, failure to conduct interviews at a convenient time and place for the complainant (see below) may discourage some complainants. As already noted, the LAPD has been found to actively discourage potential complainants. The San Jose Independent Police Auditor emphasizes that "The manner and tone used in intaking complaints is critical in instilling confidence in the objectivity and integrity of the IA and IPA offices. The objective is to instill credibility and demonstrate responsiveness to assure citizens that their grievances (real or imagined) are welcomed and will be taken seriously. Be sensitive to the message you send through your body language."[283]

Ensuring that a complaint process is receptive to complainants and that withdrawn complaints reflect the legitimate wishes of complainants requires periodic audits by some outside agency. The police auditors discussed in Chapter Six routinely do this as a part of their mission. Various options for evaluating the complaint process are discussed later in this chapter.

Police Officer Cooperation With Investigations

For decades, experts on police misconduct have argued that the greatest single obstacle to investigating alleged misconduct incidents and achieving accountability is the refusal of other officers to cooperate with investigations. This includes investigations related to citizen complaints, internal police department investigations, and criminal investigations by prosecutors. The so-called "code of silence" involves four distinct actions: not reporting misconduct by other officers, falsely claiming not to have seen the events in question, actively lying to investigators, and colluding with other officers to create a cover story.[284]

Although the code of silence has long been recognized as a serious problem, only recently have police departments begun to develop policies and programs to overcome it. The first step that has been taken involves adopting a formal policy stating that it is the duty of officers to cooperate fully and truthfully with investigations. The ordinance creating the Boise Community Ombudsman requires that "Officers/employees shall, as a condition of their employment, truthfully and completely answer all questions specifically directed and related to the scope of employment and operations of Boise City that may be asked of them by any investigator or supervisor acting on behalf of the Office of the Community Ombudsman."[285] As discussed in Chapter Three, an increasing number of police departments have adopted formal policies explicitly stating that officers have a duty to report misconduct by other officers.

With respect to collusion among officers to create an agreed-on cover story, two remedies have recently emerged. First, in several jurisdictions staff members of independent agencies immediately "roll out" to all shooting or serious use of force incidents (this procedure is discussed below and in

Chapter Three). Second, the consent decree in Los Angeles requires that when two or more officers are involved in an incident, they are to be immediately separated to prevent collusion.[286]

Departmental Cooperation

In cities and counties where an external citizen oversight agency investigates citizen complaints, the police or sheriff's departments are required to cooperate with investigations as a matter of law. Yet some have been aggressive in not cooperating and even obstructing investigations. In 2001, for example, the San Francisco Office of Citizen Complaints (OCC) encountered 99 incidents of officers failing to cooperate with investigations as required by law. These cases were forwarded to the San Francisco Police Department, but it sustained only 39 of them, and only 7 of those resulted in the officer receiving a written reprimand.[287]

The same 2003 report by the San Francisco OCC found a number of flagrant instances of the police department deliberately stalling or interfering with OCC investigations. These practices included changing policies on the release of documents without notifying the OCC, thereby creating delays; failing to respond to legitimate requests for documents; withholding documents in some cases for more than a year; refusing to provide photographs of officers and mug shots of complainants; and delaying release of documents even after the Department has agreed to release them.[288] Systematic non-cooperation or obstruction of this sort undermines the investigations of complaints and the entire process of accountability.

Interviews at Convenient and Comfortable Locations

Pursuing a complaint is a time-consuming process for the average citizen, and this may account for a certain number of withdrawn complaints. Officers are paid for being interviewed because it is an official responsibility, but complainants must take time off from either work or family. The lack of convenient locations for interviews only discourages complainants even further. The very high attrition rate among complainants in New York City may be due to the fact that complainants must travel to the CCRB office at the very tip of Manhattan— a considerable journey for someone in Brooklyn, Queens, or the Bronx.

Some agencies have taken the obvious steps to overcome this problem. The Washington, DC, OCCR has staff vehicles that allow its investigators to visit incident scenes and meet complainants and witnesses at convenient locations. The New Jersey consent decree, meanwhile, requires the State Police to

arrange a convenient time and place, including by telephone (or TDD), to interview civilians for misconduct investigations [and] reasonably accommodate civilians' circumstances to facilitate the progress of an investigation. This may include holding an interview at a location other than a State office or at a time other than regular business hours.[289]

Interview locations also need to be comfortable for complainants and witnesses. The San Jose IPA found that the Professional Standards and Conduct Unit provided only one cramped office with two investigators' desks for this purpose. If the other investigator were on the phone, it would be disruptive to the interview, and if the complainant or witness wanted to speak privately, the other investigator would have to leave the room.[290]

Ensuring Thorough and Fair Investigations

Although there is universal agreement that complaint investigations should be thorough and fair, it is only comparatively recently that formal standards have been developed for defining and ensuring thoroughness and fairness. The Omaha Public Safety Auditor, for example, regards an investigation as fair if "all witnesses [are] treated the same;" complete when "all pertinent evidence [is] in the file . . .;" thorough if "all issues [were] identified and investigated;" and objective if "the interview and investigation [were] conducted free of bias."[291] Boxes 4.1 and 4.2 present the criteria used by the San Jose Independent Auditor in evaluating the thoroughness of both investigations and interviews.

The failure of both police internal affairs units and some citizen oversight agencies to meet these standards of thoroughness and fairness have been documented in numerous reports. The court-appointed monitor in Pittsburgh, for example, found an excessive force complaint in which the Office of Municipal Investigations (OMI) failed to examine the medical records of treatment received by the complainant (or at least failed to document that it reviewed those records). In addition, the OMI conducted no canvas of the incident scene to locate potential witnesses despite the fact that the incident occurred at a crowded bar.[292]

The following issues are particularly important to ensuring thorough and fair investigations.

Locating and Interviewing Witnesses

Failure to locate and interview witnesses is a technique that is likely to ensure that a complaint will not be sustained. Failure to take the necessary steps is endemic in police complaint procedures and an indication of disinterest in complaints. A recent report on the complaint process in Albuquerque, for example, found one use of force case in which there were several potential witnesses to the incident who were not interviewed. In some other cases "witnesses were either not asked the right questions or not asked any questions at all."[293]

The Los Angeles consent decree directs the department to take active steps by "canvassing the scene to locate witnesses where appropriate, with the burden for such collection on the LAPD, [and] not the complainant." Investigators with the Oakland, California, Citizens Police Review Board (CPRB) printed flyers that they would post in the neighborhood, asking if there were witnesses to an incident under investigation.[294]

Obtaining Medical Records

In cases in which the allegation involves serious use of force, the complainant may have received medical treatment. A thorough investigation includes an effort to determine if this was the case and, if so, an effort to obtain and examine any such records.

Avoiding Conflicts of Interest

There are always potential conflicts of interest between investigators and officers being investigated. The conflicts include the obvious family or marital ties as well as more subtle ones arising from previous working relationships, close friendships, and past conflicts.

The San Jose Independent Police Auditor was perhaps the first agency to address this issue, with a formal policy stating that "In order to avoid bias, IA investigators are required to advise the Unit Commander of conflicts due to prior friendships, frequent interaction or adverse contacts with the complainant."[295] More recently, the consent decree with the New Jersey State Police adopted a similar policy stating that "The State shall prohibit any state trooper who has a conflict of interest related to a pending misconduct investigation from participating in any way in the conduct or review of that investigation."[296]

Prohibiting Leading or Hostile Questions

Complaint investigations are biased if investigators ask leading questions of police officers or put words in their mouths or provide legal justifications for their actions. In Portland, the director of the auditor's office noticed that in cases in which an officer would hesitate when faced with a difficult question, internal affairs investigators would suggest an exculpatory answer.

The Cincinnati consent decree prohibits investigators from "improperly asking officers or other witnesses leading questions that improperly suggest legal justifications for the officer's conduct when such questions are contrary to appropriate law enforcement techniques."

Probing Inconsistencies

Police internal affairs investigators have often ignored misconduct by simply failing to probe inconsistencies in statements by officers (either inconsistencies by one officer or among the statements by two or more officers). Along the same lines, investigators have often accepted at face value patently ludicrous statements by officers. The San Jose IPA Policy and Procedures addresses this issue by requiring that "Inconsistent statements of material issues should be analyzed. This analysis should be applied to both citizen and police witnesses/subject officer's statements."[297]

Judging the Credibility of Officers and Complainants

A particularly serious pattern of bias in complaint investigations has been the practice of automatically giving credence to the officer's testimony and discrediting the statements of the complainant. This has occurred even in situations in which the officer's testimony is patently ludicrous. The Office of Independent Review (OIR) in the Los Angeles Sheriff's Department, for example, found that a "good guy" principle was inappropriately used to mitigate discipline in some internal investigations.[298]

A manual on *Investigating Workplace Harassment,* published by the Society for Human Resource Management (SHRM), warns that "Determining the truth from two or more conflicting stories can be a very difficult task." Citizen complaints against police officers inevitably involve conflicting stories. They are typically "he said/he said" situations. The SHRM further advises that, despite what many people think, it is very difficult to make credibility determinations on the basis of peoples' demeanor. The truthful witness may exhibit great nervousness, whereas someone lying may exude extreme confidence and composure.[299] For this reason, corroborating evidence is extremely important, which in turn heightens the importance of locating and interviewing potential witnesses and gathering medical evidence where the complainant was injured.

Traditionally, police internal investigations have automatically accepted the statements of the officer over those of the complainant or witnesses, in some cases despite the fact that the officer's statement is filled with inconsistencies or simply not believable. The Cincinnati consent decree requires that "There will be no automatic preference for an officer's statement over a non-officer's statement, nor will [investigators] completely disregard a witness's statement merely because the witness has some connection to the complainant."[300] The Los Angeles consent decree contains a similar requirement.[301] The consent decree with the Riverside, California, Police Department goes one step further, requiring investigators to provide an explicit rationale for their credibility judgments. Investigators are also directed to take into account an officer's complaint history in making a credibility determination.[302]

Investigating Collateral Misconduct

In the course of some—perhaps even many—complaint investigations, investigators discover additional officer misconduct that is not part of the formal complaint. This is understandable. A citizen may file a complaint

Table 4.1   Collateral Misconduct, San Francisco OCC and
New York City CCRB

	Complaints Received	Total Allegations	Allegations per Complaint
San Francisco, OCC, 2001	999	4321	4.3
New York City CCRB, 2002	4512	7216	1.6

Sources: San Francisco, Office of Citizen Complaints, 2001 Annual Report (San Francisco: Office of Citizen Complaints, 2002); New York City Civilian Complaint Review Board, Status Report, January–June 2003 (New York: Civilian Complaint Review Board, 2003).

about excessive force but not realize that the officer also violated two other department policies. These additional acts of misconduct are referred to as *collateral misconduct* or, colloquially, misconduct "outside the four corners" of the formal complaint. Many agencies simply ignore collateral misconduct, claiming they are only responsible for investigating what the complainant alleged. Ignoring collateral misconduct is a way of both protecting officers and limiting the investigative workload.

Pursuing collateral misconduct is a hallmark of a police department or citizen oversight agency that is serious about accountability. The experience of the San Francisco Office of Citizen Complaints (OCC) is revealing on this issue. The average number of allegations per complaint rose from 1.88–2.81 in the 1989–1996 period to an average of 4.29–4.78 between 1997 and 2000).[303] This increase may have been a direct result of an increase in the staffing in the OCC in 1997 (see below) and the leadership of Mary Dunlap, appointed director of the OCC in 1997. By comparison, the New York City CCRB, which has historically had staffing problems, reported only 1.6 allegations per complaint in the first half of 2003.[304] The net result is the San Francisco OCC is investigating far more allegations of officer misconduct than is the NYC CCRB.

Following the lead of the OCC, the Los Angeles consent decree specifies that if an investigator believes that misconduct may have occurred "other than that alleged by the complainant, the alleged victim of misconduct, or the triggering item or report, the investigating officer must notify a supervisor, and an additional Complaint Form 1.28 investigation of the additional misconduct issue shall be conducted."[305]

The San Jose IPA, meanwhile, requires that "Intake investigators have a duty to identify all allegations raised in the complainant's statements regardless of whether the complainant is or is not complaining about the particular conduct." To illustrate the point, it provided the following example:[306]

> For example, the complainant alleges that when he was handcuffed, the cuffs were too tight causing bruising to his wrists. The officer had not double locked the cuffs and ignored the request to loosen them. The complainant asked for the officer's name and badge number and threatened to file a complaint. The officer refused to provide identification. The citizen is complaining only about the injuries to his wrists. The SJPD policy is that officers provide their name and badge numbers to any citizen requesting identification. The appropriate allegations would be Unnecessary Force and Improper Procedure (not providing identification).

Box 4.1   San Jose Independent Police Auditor Checklist to Assess
                Thoroughness of Investigations

Were all the identified witnesses interviewed? If not, why? The Auditor may send a request to conduct the missed interviews or have IA explain what efforts were made to interview these witnesses.

What efforts were made by the IA investigators to find additional witnesses?

Was a neighborhood canvas conducted? Were leads from the complainant or other witnesses developed?

Did the investigation include any photographs or diagrams?

Was the IA investigator objective in writing the final comprehensive report?

Were consistencies and inconsistencies between civilian witnesses pointed out? Were consistencies and inconsistencies between police officers also pointed out?

Were the facts as represented in the IA reports consistent with the contents of the taped interviews?

From San Jose, IPA, Policy and Procedures. www.ci.san-jose.ca.us/ipa/home.html.

Box 4.2   San Jose Independent Police Auditor Checklist to Evaluate
                   the Quality of Interviews

Did the IA investigator encourage the witness to feel at ease prior to beginning the interview?

Was the witness allowed to give an uninterrupted statement? Was the witness allowed to explain his or her answers?

Did the IA investigator interject his or her own personal opinions or rationalize the officer's behavior?

Was the IA investigator discourteous or confrontational?

Were all relevant issues covered in the interview?

Was there any discussion with the witness that was not recorded?

Were the police and the civilian witnesses admonished not to discuss the case with other witnesses or officers?

Were the questions leading or open-ended, and were follow-up questions asked?

Was the IA investigator's demeanor, intonation of voice different toward citizens than officers?

Was applicable policy or law covered in the officer's interview?

Was the overall manner of conducting the interview objective?

Tape-Recording Interviews

One of the most important new techniques for ensuring the quality of complaint investigations is the practice of tape-recording all interviews with officers, complainants, and witnesses.[307] The tapes can then be reviewed by either supervisors or an external oversight agency to identify inadequacies that need to be corrected. This might involve additional investigation of the case in question or staff training to improve interview techniques. The tapes can also be used for personnel evaluations of investigators, including recommendations for transfer or termination.

Tape recording, first developed among citizen oversight agencies, has spread to police departments, and is required by some consent decrees.[308] The process of reviewing the tapes of investigations is one of the important functions of the police auditor form of citizen oversight, and one of the reasons why some experts on the subject (including the author of this book) argue that the auditor model of oversight is a better alternative than the traditional form of civilian review of the police. (The auditor model is discussed in detail in Chapter Six). Merely transferring responsibility for investigating complaints to a nominally independent agency does not in itself ensure that complaint investigations will be of high quality. The Portland, Oregon, auditor reviewed tapes and regularly identified cases that needed to be reinvestigated and identified investigators who needed retraining or better supervision.[309]

Reopening Investigations

The tape-recording of interviews is only one device for reviewing the quality of investigations. Other techniques include having an independent observer at interviews and having supervisors carefully reviewing investigative case files. The San Jose IPA has the authority to observe interviews and review both tapes and case files, and the effectiveness of this process is indicated by the fact that in the first half of 2003, the IPA requested further investigative action in 31% of the cases it reviewed. In one case, in which a person stopped for a traffic violation claimed to have been stopped because of race, the complainant demanded to speak to the officer's supervisor. The officer said he was only going to give a warning, but then wrote a citation for failure to signal a lane change. The driver was ultimately acquitted of the traffic citation. IPA reviewed the officer's enforcement records, but could find no other citations for this offense, suggesting that this was a retaliatory action against a citizen who asked to speak to a supervisor.[310]

Disposition Separate From Investigation

It has also been recommended that the decision related to the disposition of complaints should be separate from the investigation of complaints. Separating the two helps to ensure that the investigation is thorough and not cut short by a premature determination that the complaint will not be sustained.[311]

The investigation should generate a formal investigative report that summarizes the facts of the case. One problem that has occurred in some police departments is that important facts are omitted—probably deliberately—from the investigative report, with the result that the complaint is not sustained. This problem only highlights the importance of maintaining high professional standards for investigators. In one of her first actions, Teresa Guererro-Daley, the San Jose Independent Police Auditor, found that the log in complaint investigation files "contained only scant, handwritten notes that were very difficult to read or audit." As a result, some complaints were improperly placed in a less serious category. On her recommendation, a standard form was developed and all information was typed and entered into a central database.[312] Complete, factually correct, and legible investigative reports facilitate the process of auditing complaint files and determining whether dispositions are properly based on the facts of the case. Incomplete and sloppy files, by the same token, are indicators of unprofessional investigative practices.

Standards for Disposition of Complaints

Citizen complaints are resolved or disposed of in one of four categories: *Unfounded* (there is no evidence that the alleged misconduct occurred); *Not Sustained* (the misconduct may have occurred, but there is insufficient evidence to resolve the issue either way); *Exonerated* (the alleged conduct occurred but was proper); and *Sustained* (the officer committed the misconduct as alleged). Until fairly recently, complaint review procedures had no clearly articulated standards for deciding how to resolve a complaint investigation. For all practical purposes, police internal affairs units sustained only those complaints in which the evidence was overwhelmingly against the officer.

By the 1990s, citizen oversight agencies began developing disposition standards. Among oversight professionals the preferred standard is the *preponderance of the evidence.* This standard is defined as meaning that "it is more likely than not that the alleged action occurred."[313] This is a far lower standard than the *proof beyond a reasonable doubt* standard used in the criminal process and also lower than the *clear and convincing evidence* standard used by some complaint review procedures. The preponderance of the evidence standard has been adopted in most of the consent decrees negotiated by the Department of Justice, including the decrees of the New Jersey State Police and the Los Angeles Police Department.[314]

Dispositions Related to the Facts

Another problem with dispositions is that some are not rationally related to the facts developed in the investigation. The monitor in Pittsburgh, for example, found "illogical leaps to unfound cases" in 8 of 32 cases that were examined. Several cases alleging improper handcuffing and excessive force, for example, were unfounded because the complainants were "found or pled guilty to an arrestable offense."[315]

Feedback to Complainants and Officers

Lack of feedback about the status of complaints is a major problem among both complainants and police officers. At best, complainants generally receive a form letter indicating the final disposition of their complaint but with little explanatory detail. Complainants are often alienated by the combined impact of delays in the completion of investigations, the lack of notice about the delays, and the lack of explanation about the disposition. Police officers are often not treated much better, receiving no updates about the status of an investigation or a letter about the final disposition. This only creates morale problems among the rank and file.

In Philadelphia, the Integrity and Accountability Office found that the department had stopped sending letters to complainants regarding the results of misconduct hearings, in direct violation of an executive order requiring such letters. "We could not identify any policy or system in place for ensuring that complainant's [sic] are notified of the results of their complaints in a timely and accurate fashion."[316]

The Washington, DC, Office of Citizen Complaint Review (OCCR) has taken an extra step in providing feedback and informing the public by making its decisions available to the public on its website. Philip K. Eure, OCCR's executive director, explains that this process will help both citizens and officers "better understand how the process works and how other complaints have been resolved."[317]

Staffing, Managing, and Evaluating the Complaint Review System

Staffing and Resources

It should go without saying that to function effectively a complaint review procedure needs sufficient staff and resources to handle the complaint workload it receives. Unfortunately, it needs to be said. Lack of appropriate staffing has been a major problem for both police internal affairs units and citizen oversight agencies. In a critical report on the New York City CCRB, the New York Civil Liberties Union (NYCLU) found that "Virtually all investigator hires were entry-level employees, who were overwhelmed by a large case back log and a manual record-keeping system."[318] The police auditor in Portland, Oregon, found that the excessive delays in investigating complaints by the police department were largely the result of a lack of sufficient investigators in the internal affairs units.[319]

There are, however, no professional standards regarding the appropriate number of investigators given the size of a police department. Both the CALEA accreditation standards and the IACP concept paper on investigating misconduct are silent on this issue.[320] In addition, there is no research on how much investigative effort is required to investigate the typical complaint. Anecdotal evidence indicates that a triage formula should be used, as the more serious complaints (e.g., a use of force complaint with multiple allegations, several officers, and a number of witnesses) require a great deal of time and effort, whereas many complaints require very little (especially those with no witnesses or other forensic evidence).

The only existing standards apply to the San Francisco Office of Citizen Complaints. A 1997 ordinance, enacted by referendum, requires the OCC to have one complaint investigator for every 150 sworn officers in the San Francisco Police Department. With a police department of about 2,300–2,400 sworn officers, the OCC had about 16 staff investigators for most of 2000.[321] The OCC has also taken the lead in reporting workload data. In 2001, for example, it reported that its case closure rate for the year "was an average of 7.0 cases per investigator, with average monthly availability of 13.5 investigators."[322] The settlement agreement with the Oakland Police Department addresses this issue, requiring that "If IAD experiences an unusual proliferation of cases and/or workload, IAD staffing shall be increased to maintain timeliness standards."[323]

With respect to police internal affairs units, determining the appropriate number of investigators is a little more difficult because in many departments IA investigators perform tasks other than investigate citizen complaints. All investigate internally generated complaints against officers, many conduct proactive investigations, and some are responsible for conducting background investigations of applicants.

An Investigation Policy and Procedure Manual

A well-managed complaint review procedure needs a comprehensive policy and procedure manual that includes, among other things, specific directives addressing the various issues previously covered in this chapter (e.g., locating and interviewing witnesses, not asking leading questions, judging witness credibility, etc.). After sitting in on more than 100 police officer interviews, the San Jose Independent Police Auditor observed that "the quality of interviews differs with each investigator." She then developed a standard format to ensure consistency and thoroughness.[324]

Although there has been no systematic survey of the subject, anecdotal evidence suggests that police internal affairs units have not developed comprehensive sets of policies and procedures. Recently, however, several of the better managed citizen oversight agencies have developed good sets of policies and procedures, and much of the material presented in this chapter is taken from them (especially the San Jose IPA, the Boise Community Ombudsman, and the Washington, DC, OCCR).[325] These local agency standards, however, have never been codified and adopted by any professional association as recommended standards.

Training for Investigators

A policy and procedure manual, in turn, becomes the basis for investigator training. Lack of consistency among investigators is a problem. The Department of Justice letter to the Portland, Maine, police department, for example, found that the sergeant assigned part-time to complaint investigations "had no prior investigative training before being assigned to the Internal Affairs Unit," and that "the lieutenants and Shift Commanders charged with conducting intake and investigating informal complaints have not received any sort of training or guidance on complaint investigation."[326] In truth, there has been very little research on police internal affairs units and we have only anecdotal evidence on this and many other issues.

The police have traditionally assumed that experience and training in criminal investigation adequately prepares officers for assignment to internal affairs. This view, however, fails to take into account the fact that investigating a citizen complaint against a fellow officer is a very different kind of task. It calls for not making prejudgments about the credibility of either the complainant or one's fellow officer. The failure to provide special training in this area accounts for the fact that internal affairs units have a bad history of asking leading questions of officers and automatically questioning the credibility of complainants.

The Special Nature of Citizen Complaints Against Police Officers

Jayson Wechter, a staff investigator with the San Francisco OCC, argues that investigating citizen complaints is very different from investigating other kinds of cases. Wechter speaks from a wide experience as an investigator for prosecutors, public defenders, and attorneys handling various kinds of civil suits. Complaints against police officers are different because the police are different. They occupy a special place in society, with the awesome power to take people's lives and their liberty. And in this country, popular culture has surrounded police officers with a special aura. Prosecutors long ago learned that it is extremely difficult to obtain judgments against police officers, for the simple reason that judges and juries almost automatically defer to their authority.[327]

Complaint investigators are influenced by the same aura of police authority. In passing, it is worth noting that in some mediation agencies, some mediators decline to accept cases involving complaints against police officers because they realize they are favorably inclined toward officers, whereas some other mediators decline because they realize they have deep-seated biases against officers.[328] Police officers, for their part, are skilled at manipulating this aura to their advantage. And in many cases the complainant is not a model citizen. Complaint investigators need to learn the habit of challenging this authority, albeit in a neutral manner. In short, special training is needed to prepare investigators for the very special task of investigating complaints against police officers.

Ensuring Timely Investigations

The failure to complete complaint investigations in a timely fashion is a pervasive national problem. The Oakland monitor in 2004 found "systemic delays at nearly every stage of the process."[329] To invoke an old cliché, justice delayed is justice denied, in this case for both complainants and the police officers who are the subject of complaints. Moreover, some citizen oversight agencies have been as guilty of unacceptable delays as have been police internal affairs units. The old Washington, DC, CCRB, which was abolished in 1997, took as long as 3 years to complete some investigations.[330]

When the San Jose IPA began operating, it found that about 25% of all cases took longer than a year to complete; in addition, some complaints were not even classified for 6 months. To correct this problem, the IPA established a set of formal timelines: All complaints must be classified and assigned to an investigator within 30 days; all investigations of Class I complaints must be completed in 180 days; and all cases were to be closed in 365 days. As an external auditing agency, the IPA then monitored the police department's success in meeting these deadlines.[331] Following these examples, some consent decrees impose timelines for the completion of investigations. The New Jersey State Police must complete misconduct investigations within 120 days.[332]

Box 4.3   Timelines for Complaint Investigations Evaluating the Complaint Process

30 Days	Classification of all complaints
180 Days	Complete investigations of all Class I Use of Force Complaints
365 Days	Complete investigations of all other complaints

From *IPA 2003 Report: A Comprehensive Ten Year Overview,* by San Jose Independent Police Auditor, 2004, San Jose, CA: Independent Police Auditor, p. 25.

The citizen complaint process is designed to hold officers accountable for their conduct. Unfortunately, insufficient attention has been given to holding complaint procedures themselves accountable for their performance. Part of the problem is that, as already mentioned, there are no professional standards for complaint procedures, and thus no standards by which to measure them.

Surveying Complainants and Officers

The most notable exception to this rule was the Quality Service Audit (QSA) developed by the now-defunct Minneapolis Civilian Review Authority (CRA). The QSA provided both complainants and officers a mail-in form allowing them to comment on their experience with the CRA. (In the interest of full disclosure, the author of this book helped the CRA design the QSA.)

The results were extremely positive: 71% of the complainants and 91% of the officers felt they had been treated with respect by the CRA staff; 76% of the citizens and 90% of the officers felt they had been able to tell their "side of the story."[333] These responses contrasted sharply with the very negative results of an earlier survey of complainants and officers who had contact with the New York City Civilian Complaint Review Board.[334] Almost all attempts to survey complainants, however, have encountered the problem of very low response rates.

Auditing the Complaint Process

In addition to customer satisfaction surveys, complaint procedures should be subject to regular and thorough audits to ensure that complaints are being investigated in a thorough and fair manner and that the process meets other standards. Experience indicates that even complaint procedures under intense public scrutiny fail to meet minimal standards of performance. The court-appointed monitor in Pittsburgh, for example, found that of 35 completed investigations in the OMI between November 16, 2001 and February 15, 2002, five were returned for further investigation. The monitor concluded that this failure rate of 14.3% was not adequate and declared the OMI out of compliance with the terms of the consent decree.[335] In 2003, the court appointed monitor in Los Angeles found that the LAPD was failing to provide proper responses to people inquiring about how to file a complaint.[336] The Pittsburgh and Los Angeles examples are particularly noteworthy because both police departments were under consent decrees in federal courts and, as a consequence, were also under intense public scrutiny. Nonetheless, both failed to comply with required mandates.

The consent decree involving the Riverside, California, Police Department calls for random audits of the complaint process at least three times a year.[337] It does not specify how these audits are to be conducted, but several different ways are obvious. First, regular surveys of both complainants and officers, as discussed above, can be conducted. Second, testers can be hired to visit police stations to inquire about how to file a complaint in order to determine whether they receive accurate information in a courteous manner. The ACLU of Northern California conducted such a survey of the Oakland Police Department and found it extremely deficient.[338] Third, regular audits of complaint files can be conducted to determine whether investigations are thorough, fair, and timely.

These audits can and should include reviews of the audio tape recordings of interviews, as recommended earlier. Several police auditor agencies, which are described in Chapter Six, undertake these activities.

The Sustain Rate as a Performance Indicator

The sustain rate, or the percentage of complaints resolved in the complainants' favor, has traditionally been used by community groups as a performance measure for police internal affairs units. And they have cited fact that only about 10% of all complaints are sustained as evidence that the police do not conduct thorough or fair investigations. The sustain rate, however, is not an appropriate performance measure. Complaints against officers are inherently difficult to sustain, usually because there are no witnesses or forensic evidence. And in fact, citizen oversight agencies do not sustain significantly higher rates of complaints than police internal affairs units. For these and other reasons, the sustain rate is not a valid measure of the effectiveness of a complaint review process. Instead, agencies should be evaluated in terms of their compliance with the best national standards for the crucial policies and procedures discussed in this chapter.[339]

CONCLUSION

Citizen complaints against police officers are an important aspect of police accountability. Citizens have a right to express their dissatisfaction with any government agency and receive a thorough and fair hearing of their complaint. Complaints, meanwhile, are a valuable form of information for police management, an indicator of a problem or problems that need to be corrected. For this reason, they are incorporated into all early intervention systems designed to identify officers with recurring performance problems (see Chapter Five). Until fairly recently, citizen complaint procedures have failed to serve these needs. A few citizen oversight agencies have developed meaningful standards for ensuring thorough and fair investigations, and an increasing number of police departments have established more open and accessible complaint procedures. The major challenge for the future involves developing national standards for complaint procedures reflecting the issues discussed in this chapter.

EARLY INTERVENTION SYSTEMS

he dirty little secret in policing is not just that some officers repeatedly engage in misconduct but that other officers know who they are. Historically, police departments always had their "problem" officers but failed to take effective action toward them. The evidence documenting the existence of the officers with performance problems is substantial. A 1981 report by the U.S. Civil Rights Commission, *Who is Guarding the Guardians?*, provided the first documentation (Table 5.1). In the Houston Police Department, one officer had received a total of 12 complaints and two others had 11 complaints over a 2-year period. By comparison, 298 officers had received only 2 complaints. The officers who had received 5 or more complaints during this period represented only 12% of all officers receiving complaints, but they accounted for 41% of all the complaints.[340]

The Houston data were confirmed by subsequent reports on other departments. The Christopher Commission found 44 officers "problem officers" in the Los Angeles Police Department and commented that they were "readily identifiable" on the basis of existing department records on citizen complaints and use of force.[341] A *Boston Globe* investigative report found two officers who had accumulated 24 complaints in the past 10 years; one had received eight complaints in 1990 alone, and the other officer had received seven that year. The Kolts Commission report on the Los Angeles Sheriff's Department found 17 deputies who were responsible for 22 civil law suits and resulted in settlements costing the county about $32.2 million. The Kolts report commented that the LASD had "failed to deal with officers who have readily identifiable patterns of excessive force incidents on their records."[342]

Table 5.1   Citizen Complaints, Houston
                   Police Officers, 1977–1979

Number of Complaints	Number of Officers
2	298
3	136
4	70
5	33
6	18
7	8
8	6
9	1
10	0
11	2
12	1
	573 Total

Note. From *Who is Guarding the Guardians? A Report on Police Practices* (p. 166), by the U.S. Commission on Civil Rights, 1981, Washington, DC: Government Printing Office.

In response to its findings, the U.S. Civil Rights Commission recommended that "A system should be devised in each department to assist officials in early identification of violence-prone officers."[343] Early intervention (EI) systems have emerged as a tool for identifying officers with recurring performance problems and correcting their performance. EI systems are now a recognized best practice in policing.[344] A Vera Institute evaluation of the implementation of the Pittsburgh consent decree commented that "the early warning [i.e. intervention] system is the centerpiece of the Police Bureau's reforms in response to the consent decree."[345]

THE CONCEPT OF EI SYSTEMS

EI systems are the centerpiece or linchpin of the new police accountability because they pull together other key elements of a department's accountability process, including primarily use of force reports (Chapter Three) and citizen complaint data (Chapter Four). A centralized database on officer performance allows commanders to identify those officers whose performance records raise concerns. These officers are then referred to some kind of intervention, typically counseling or retraining, designed to improve their performance. The performance data included in EI systems can feature anywhere from five to more than twenty items.[346]

Although EI systems originated as a tool for controlling police use of excessive force, experts now recognize that they have the capacity to identify a wide range of performance issues. They can, for example, identify officers with exemplary performance, hold sergeants accountable for the performance of officers under their command, and provide reliable systematic data on the performance of the department as a whole. EI systems can be used, for example, to analyze traffic enforcement data for the purpose of addressing the highly sensitive issue of racial profiling (see below). A 1989 report by the International Association of Chiefs of Police (IACP) recognized the many potential applications, explaining that EI systems are "a proactive management tool useful for identifying a wide range of problems [and] *not just a system to focus on problem officers.*"[347]

EI Systems in Action: Two Case Studies

In one large police department a police officer was flagged by the EI system because of a high number of use of force incidents. The counseling session with the officer revealed that she had a great fear of being struck in the face, and as a consequence was not properly taking control of encounters with citizens. After losing control over the person or persons, she would then have to use force to reassert control. Her supervisor referred her to the training unit, where she was instructed in tactics that would allow her to protect herself while maintaining control of encounters with the potential for conflict. As a result, her use of force incidents declined dramatically.[348]

In another large department, a patrol officer was identified by the EI system because of a series of use of force incidents. During the intervention session the officer's supervisor discovered that he was having severe personal financial problems. The supervisor recommended professional financial consulting, the officer followed this advice, and his performance improved significantly.

EI systems incorporate key elements of the new police accountability: the collection and analysis of systematic data on officer performance, responses to performance problems that reach deep into the organization and address specific problems, and a focus on organizational change. With respect to the latter issue, EI systems represent a proactive approach to accountability: a systematic organizational effort to identify problems before they escalate into more serious issues.

To function effectively, an EI system depends on the full development of two other elements of the new accountability: use of force reports and citizen complaints. As a consequence, it is imperative that a department have a comprehensive use of force reporting system as described in Chapter Three and an open and accessible citizen complaint process as described in Chapter Four. If either or both of those procedures fail to operate properly, the data in the EI system will not accurately reflect officer performance. In short, an EI system is not a free-standing tool that can significantly change a department by itself. It requires major improvements in other policies and procedures in a department.

The relationship of an EI system to organizational change is twofold. First, the development of an EI system itself represents an important change: a concerted effort to monitor officer performance. Second, as is explained later in this chapter, the long-term impact of an EI system has the potential for changing standards of accountability and ultimately the organizational culture of a police department. The experience with EI systems to date, however, clearly indicates that the implementation and maintenance of an EI system is an extremely difficult challenge for a police department. EI systems are extremely complex administrative mechanisms that require very close ongoing attention from the high-level commanders. Even in the relatively short history of EI systems, there are already a number of stories of unsuccessful programs.

A Note on Terminology

This book uses the term *early intervention* rather than *early warning,* which was used when the concept first originated.[349] The term *early warning* has a negative connotation, suggesting that the system is primarily oriented toward discipline. One department with a comprehensive EI system found through interviews with officers that they did not like the phrase "early warning" because of its "big brother" connotation. EI systems are evolving in the direction of more comprehensive personnel assessment systems that examine more than just use of force issues. The Los Angeles Sheriff's Department calls its system the *Personnel Performance Index* (PPI); the Phoenix Police Department calls its system the *Personnel Assessment System* (PAS); whereas in Cincinnati it is known as the *Risk Management System* (RMS).

This book also uses the term *officers with performance problems* rather than the commonly used term *problem officers*. The latter terms unfairly labels officers and suggests that their performance cannot change. The term *officers with performance problems* focuses on behavior without labeling an officer and conveys the message that performance can improve.

Early Intervention Versus the Formal Discipline System

Early intervention systems are separate from a police department's formal discipline system. The discipline system involves official actions toward officers in response to sustained allegations of misconduct. It is reactive, initiated only after misconduct has been confirmed. The resulting disciplinary actions are documented in an officer's official file.

EI systems, on the other hand, are designed to help officers improve their performance, and the actions taken toward an officer are informal, flexible, and confidential. Informality is designed to permit the department to address a wide range of issues, some of which might involve personal matters that are affecting an officer's performance. No formal record of the content of an intervention—what an officer says, any recommendation for professional counseling—is maintained. EI systems are also proactive, seeking to address performance problems before they become a matter for formal disciplinary action. One important aspect of EI systems is their capacity to identify patterns reflecting multiple performance categories—for example, a pattern of high speed pursuits (none of which resulted in disciplinary actions) *and* excessive use of sick leave, a combination that suggests the officer may be under stress due to off-the-job problems.

An EI system and the discipline system can respond simultaneously— although separately—to the same incidents. An officer with four use of force incidents might be disciplined for one of those incidents while the EI system examines the broader patterns and seeks to understand and respond to the underlying causes.

One of the traditional failures of police personnel practices is that they are oriented toward punishing rather than helping employees. Police departments have been characterized as punishment-oriented bureaucracies, with innumerable rules and regulations that can be used to punish an officer, but with few procedures for either rewarding good conduct or helping officers with problems.[350] Apart from employee assistance programs (EAP) designed to address substance abuse or family problems, police departments have done relatively little in a formal way to correct problem behavior.[351]

Traditional performance review systems rely heavily on very general categories and subjective assessments such as "works well with people" or "demonstrates initiative."[352] They also tend to be heavily influenced by officers' reputations, which may be invalid or not reflective of current performance. EI systems, however, can identify specific areas of performance that need correcting, such as a pattern of citizen complaints alleging rudeness, and develop a response tailored to that problem.

Contrary to what some people believe, EI systems are not devices for *predicting* officer performance on the basis of background characteristics or any other set of factors. Rather, they are retrospective performance reviews. In the past there have been several attempts to develop methodologies for predicting which applicants for police employment will perform well and which are unsuited for police work. Almost all of these efforts have attempted to correlate background characteristics with subsequent performance. None has proven to be successful.[353] An EI system makes no attempt to predict performance. It is simply a retrospective analysis of performance followed by an intervention designed to correct whatever problems have been identified.

Two Types of Early Intervention Systems

There are two basic types of EI systems. Large systems, such as the Pittsburgh Performance Assessment and Review System (PARS) and the Phoenix PAS systems, can be characterized as *comprehensive personnel assessment systems*. They collect a very wide range of data and have the capacity to address a wide range of issues, such as identifying top performing officers or investigating racial profiling. These systems require both a sophisticated technological infrastructure and an enormous amount of administrative oversight.[354]

Many other systems, such as the Miami-Dade and the Tampa, Florida, systems, are more limited in their approach and can be characterized as *performance problem systems*. They collect a smaller range of performance data and as a consequence have more limited capabilities. Being much smaller, however, they are also less expensive and do not impose the enormous administrative burdens of the more comprehensive EI systems.

THE BACKGROUND AND DEVELOPMENT OF EI SYSTEMS

The Emergence of a Concept

Early intervention (EI) systems grew out of a recognition that a few officers are responsible for a disproportionate number of any police department's
problems related to use of force and citizen complaints. The existence of the
so-called "problem officer" has been known informally for a long time. The
evidence was always anecdotal, however, and departments rarely did anything
to address the problem. Police chiefs often commented that "90 percent of our
problems are caused by 10 percent of our officers."[355] In some instances, problem officers were "buried" by giving them assignments in which they could do
little harm to the public. But these informal punishments were used arbitrarily
and also used against officers who were out of favor with the high command.
Professor Herman Goldstein, one of the leading experts on policing, was in
1977 the first person to discuss the possibility of "Identifying Officers with a
Propensity for Wrongdoing" and taking administrative action to improve their
performance.[356]

In the 1970s, several departments developed experimental programs to
identify officers involved in frequent shooting incidents.[357] In its 1981 report,
the U.S. Civil Rights Commission cited with favor programs developed in the
Oakland, New York City, and Kansas City police departments.[358] None of these
early programs appear to have survived very long, however (and in fact may
never have been fully operational). The 1984 Mollen Commission report on
corruption and violence in the New York City Police Department, for example,
offers no evidence of a functioning EI system.[359] The Oakland Police Department had no functioning EI system at the time a 2002 consent decree directed
it to develop one.[360]

The first EI programs that have survived to the present are in the Miami-
Dade and Miami police departments. The Miami-Dade Employee Identification
System (EIS) system developed in response to a series of racial incidents in the
late 1970s and early 1980s. The most controversial incident involved the fatal
beating of Arthur McDuffie, an African American insurance agent, by Miami-
Dade officers, and a riot that erupted in May 1980 after four officers were
acquitted of criminal charges for his death.[361] In response to these problems, a
local ordinance directed the Miami-Dade Police Department to develop an
Employee Profile System (EPS) to provide detailed information on each
employee. The EPS system became the basis for the EIS. At about the same

time, the city of Miami Police Department also developed an early EI program in response to racial conflict with the community.

During the 1980s, some other police departments across the country attempted to copy the Miami-Dade EIS system. A major turning point occurred with the highly publicized beating of Rodney King by Los Angeles police officers in March 1991. The Christopher Commission, appointed to investigate the LAPD following the King incident, identified 44 "problem offi-cers" with extremely high rates of complaints in the department.[362] Particularly important, the Commission pointed out that these 44 officers were "readily identifiable" through existing departmental records. The LAPD did not use these records to address the officers' performance problems, nor did it incor-porate the information into its regular performance evaluations or promotion decisions.[363] The year following the Christopher Commission report, the Kolts Commission investigating the Los Angeles County Sheriff's Department (LASD) found a similar group of 62 "problem officers" in the department and recommended the development of an EI system.[364] This led to the creation of the Personnel Performance Index (PPI), which became fully operational in 1997 and soon became widely regarded as the best EI system in the country.

The Emergence of a "Best Practice" in Policing

By the late 1990s, early identification systems emerged as a recommended "best practice" with regard to police accountability. The Department of Justice recommended EI systems in its 2001 report, *Principles for Promoting Police Integrity,*[365] and they are mandated by all of the settlements in pattern or prac-tice suits brought by the Civil Rights Division of the U.S. Department of Justice since 1997. These cases include the suits against the Pittsburgh Police Bureau, the New Jersey State Police, the Metropolitan Police Department of Washington, DC, the Los Angeles Police Department, and the Cincinnati Police Department.[366] Finally, in 2001 the Commission on Accreditation for Law Enforcement Agencies (CALEA) adopted a new standard requiring all large agencies to have an EI system.[367] Standard 35.1.15 reads:

> A comprehensive Personnel Early Warning System is an essential component of good discipline in a well-managed law enforcement agency. The early identification of potential problem employees and a menu of remedial actions can increase agency accountability and offer employees a better opportunity to meet the agency's values and mission statement.

Promulgation of the standard reflects the extent to which the concept of EI systems has permeated professional thinking in American law enforcement and been acknowledged as a best practice.

THE COMPONENTS OF AN EI SYSTEM

An EI system consists of four basic components: performance indicators, the identification and selection process, intervention, and postintervention monitoring.[368]

Performance Indicators

The performance indicators are those officer activities that are officially recorded by the department and entered into the EI system database. There is no consensus of opinion among experts about the number of performance indicators to be included in an EI system. As discussed above, there are two basic types of EI systems depending on the number of indicators used. The Pittsburgh PARS system uses 18 performance indicators, while the Phoenix PAS system utilizes 24 indicators. Other departments use as few as five indicators.

Almost all experts agree that an EI system should *not* rely on just one indicator, as many of the first EI systems did (and some still do). The initial Minneapolis early warning system, for example, used only citizen complaints.[369] There are a number of problems related to official data on citizen complaints, and they are not consistently reliable indicators of officer performance.[370] Using a broad range of indicators presents a fuller picture of an officer's overall performance.

Figure 5.1 lists the 20 indicators mandated by the consent decree covering the Oakland, California, police department.[371] In addition to use of force reports and citizen complaints, Oakland also includes sick leave usage (Item #17). Many police commanders believe that frequent use of sick leave is an indicator of possible personal problems, such as substance abuse. Resisting arrest charges filed by an officer (Item #13), meanwhile, are seen by many experts as a device by which officers cover their own use of force by charging the citizen with an offense that would justify the use of force.[372] Criminal suspects do resist arrest and even the best officers will file resisting arrest charges, but an EI system can identify a pattern where an officer is filing a much higher rate of such charges than peer officers, and where there are other indicators of potential problems.

1. All uses of force required to be reported by OPD.

2. OC spray canister check-out log (see Section V, paragraph D).

3. All police-canine deployments.

4. All officer-involved shootings and firearms discharges, both on duty and off duty.

5. All on-duty vehicle pursuits, traffic accidents, and traffic violations.

6. All citizen complaints, whether made to OPD or CPRB.

7. All civil suits and/or tort claims related to members' and employees' employment at OPD, or which contain allegations that rise to the level of a *Manual of Rules* violation.

8. Reports of a financial claim as described in Section VI, paragraph G (3).

9. All in-custody deaths and injuries.

10. The results of adjudications of all investigations related to items (1) through (9), above, and a record of all tentative and final decisions or recommendations regarding discipline, including actual discipline imposed or nondisciplinary action.

11. Commendations and awards.

12. All criminal arrests of and charges against OPD members and employees.

13. All charges of resisting or obstructing a police officer (Penal Code §§69 and 148), assault on a police officer (Penal Code §243(b)(c)), or assault-with-a-deadly-weapon on a police officer (Penal Code §245(b)(c)).

14. Assignment and rank history for each member/employee.

15. Training history for each member/employee.

16. Line-of-duty injuries.

17. Sick leave usage, particularly one-day sick leaves.

18. Report Review Notices or Case Evaluation Reports for the reporting member/employee and the approving supervisor.

19. Criminal cases dropped due to concerns with member veracity, improper searches, false arrests, etc.

20. Other supervisory observations or concerns.

Figure 5.1 EI system requirements in the Oakland (CA) consent decree. From *Allen v. City of Oakland,* Consent Decree (2003). The settlement agreement is available at www.oaklandpolice.com.

As already discussed, there are two basic types of EI systems depending on the number of indicators used. The decision about the number of indicators to use involves a trade-off between effectiveness and efficiency. A larger number of indicators increases the capability of the system to analyze the full scope of an officer's performance and identify a variety of different problems. (Figure 5.2 and the accompanying discussion below provide five examples.) At the same time, however, this imposes a very heavy administrative burden on the department involving data entry and overall system management.[373] A smaller EI system with fewer indicators has more limited capabilities but is cheaper and far easier to create and maintain.

Identification and Selection of Officers

The identification and selection of officers for intervention is an extremely complex process. In many early EI systems, identification and selection was a single, nondiscretionary decision. In the early Minneapolis system, for example, any officer receiving three complaints in a 12-month period was automatically referred to intervention.[374] Many experts, including this author, now argue that this approach is too rigid and creates a number of potential problems. First, using only one performance indicator fails to capture the full range of an officer's performance. Second, a rigid formula of three complaints in a 12-month period (or three use of force reports in a given period) does not take into account the nature and context of these incidents. There could be very legitimate reasons why an officer received that many complaints during a given reporting period. It could be the result of a rash of gang-related activity in the neighborhood that involved the officer in a number of difficult arrests that required the use of force. Third, a single department-wide formula (e.g., three complaints in a 12-month period) fails to take into account the enormous differences in officers' work environments. An officer working in a high-crime precinct will inevitably use force more often and receive more complaints than an officer working in a low-crime area. An officer in a low-crime area who receives two complaints will not be identified by the system but in fact may have serious performance problems.

The more sophisticated EI systems today treat identification and selection as two separate stages, with the result that some officers who are initially identified on the basis of the data may not be selected for intervention. An officer might have a relatively high number of problematic performance indicators, such as use of force reports and citizen complaints. A full review of that officer's performance, however, might discover legitimate explanations for the use of force reports (as explained above with regard to complaints). At the time, another officer with fewer use of force reports and citizen complaints might have a number of other indicators, such as use of sick leave and resisting arrest charges, that suggest the need for intervention. In short, *the numbers in an EI system database do not speak for themselves.* They require further inquiry to ascertain the context of an officer's work assignment and performance history.

EI systems today use different formulas for identifying and selecting officers. The Miami-Dade EIS system identifies officers who have either two or more citizen complaints or three or more use of force reports in any quarterly reporting period. The Los Angeles Sheriff's Department identifies officers who have been involved in a disproportionate number of "risk" incidents, or on the basis of a supervisor's referral. Officers who are initially identified are then subject to screening by the Performance Review Committee. The committee consists of rotating panels of three commanders that meet twice a month. The committee solicits the opinion of the officer's captain and then assigns a lieutenant to prepare of detailed Employee Profile Report (EPR; known informally as a "Blue Book"). The Performance Review Committee uses the EPR to decide whether or not to refer the officer to Performance Review. Between 1996 and 2002, a total of 1,213 employees were identified by the PPI system, but only 235 (or 19%) were placed on Performance Review. When an officer is placed on Performance Review, a Performance Plan is prepared, which might include counseling, retraining, reassignment, or a more thorough fitness for duty evaluation.[375]

As should be evident, the two-stage process in the LASD's PPI is both highly discretionary and labor intensive. This imposes a considerable administrative burden in terms of analyzing the data, undertaking a full assessment of officers' performance, and then making decisions about referral to intervention. This simply reinforces the point that an EI system is not easy to manage and requires a tremendous ongoing commitment of human resources to a department.

Peer Officer Comparisons

A two-stage identification and selection process still does not resolve the question of what criteria or *thresholds* should be use in identifying officers and then selecting some of them for intervention. A growing consensus of opinion suggests that the best approach is to use a peer officer comparison approach. Comparing officers who work the same assignment takes into account the variations in the work assignments. It is assumed that officers working in high crime areas will be more likely to use force and receive complaints than officers working in low crime areas.[376] Officers working as detectives or in a gang unit will have very different kinds of work experience than patrol officers.

The Pittsburgh PARS system uses peer officer comparisons,[377] and the Cincinnati Memorandum of Understanding directs the Risk Management System to use a similar approach. Specifically, the Cincinnati risk management system is required to report data on the "average level of activity for each data category by individual officer and by all officers in a unit. . . ."[378] One of the great advantages of the peer officer comparison approach is that it can be used to identify a variety of different performance problems. In its most sophisticated application, it can be used to identify top performing officers as well as officers who are doing little if any police work. Figure 5.2 presents several hypothetical examples of how a peer officer comparison approach can be used.

Special Application: EI Systems and Racial Profiling

As the case of Officer E in Figure 5.2 indicates, comprehensive EI systems have the capacity to address the contentious issue of racial profiling (although this is possible only with a comprehensive personnel assessment type of EI system). The problem that has bedeviled traffic stop data collection efforts is the issue of the proper benchmark for determining whether or not inappropriate racial bias exists in traffic enforcement. Most data collection efforts have used census data on the resident population data (whether state, county, city, or even precinct) even though most criminologists recognize that they are not an appropriate benchmark. Population data do not reflect the racial and ethnic composition of the driving population, much less the "at risk" driving population of law violators.[379]

A peer officer comparison approach resolves the problem of the benchmark by analyzing the EI system data to identify officers who stop a higher proportion of African American or Latino drivers than their peers. It is assumed that in a predominantly Latino neighborhood, a high percentage of all stops will involve Latino drivers, and further that the traffic stop activities of all officers working that assignment should be roughly equal. The EI system

Officer A

Indicators. Officer A had five use of force reports during one reporting period. Relative to other officers in the same unit, this was a very low figure, but his performance record indicated that he made only eight arrests during this period.

Analysis. Officer A used force in more than half of the arrests he made. The ratio of force to arrests is cause for departmental concern and probable intervention.

Discussion. Officer A represents the classic officer with a use of force problem. An intervention counseling session will seek to determine if the cause is improper tactics that can be corrected through training, or personal problems that require professional counseling.

Officer B

Indicators. Officer B had no citizen complaints or use of force reports for the reporting period. The performance data, however, also indicates that he also had made no arrests, no traffic stops, and no field stops. Further examination of his records indicate that he was working the maximum number of hours permitted for off-duty employment.

Analysis. Officer B is devoting all his energy to his off-duty employment rather than fulfilling his responsibilities to the department and will be referred for intervention.

Discussion. In the intervention counseling session the officer will be advised of this problem and instructed to engage in an acceptable level of basic police work.

Officer C

Indicators. A female driver filed a citizen complaint against Officer C alleging an inappropriate sexual advance during a traffic stop. An examination of the officer's EI file indicated a very high number of traffic stops involving females relative to peer officers.

Analysis. Officer C appears to be abusing his law enforcement powers to harass female drivers and will be referred for intervention.

Discussion. In the intervention counseling session Officer C will be presented with the data, advised that his behavior is inappropriate, and informed that he will be subject to intense supervision for the next 6 months.

(Continued)

Figure 5.2   Hypothetical examples of how a peer officer comparison approach can be used

Figure 5.2 (Continued)

Officer D

Indicators. Officer D makes a high number of arrests relative to his peer officers who work in a high crime area, yet he receives few citizen complaints compared with his peers.

Analysis. Officer D is an exemplary officer, engaging active crime-fighting work and conducting himself in a professional manner.

Discussion. The officer will be advised by his supervisor that his performance is exemplary and that a letter of commendation to this effect will be placed in his file.

Officer E

Indicators. Officer E makes roughly the same number of traffic stops as peer officers working the same precinct, a neighborhood with a significant Latino population, but stops a far higher percentage of Latino drivers than the other officers (about 60% of all stops, compared with about 40% for other officers).

Analysis. The officer's traffic stop data suggests possible bias against Latino drivers and will be referred for intervention.

Discussion. Officer E will be presented with these data at the intervention counseling session. He will be offered an opportunity to explain the disparity. If he fails to present a reasonable explanation, he will be advised of the apparent pattern of bias and informed that his performance will be subject to intense supervision over the next 6 months.

---

data can readily identify an Officer E who is stopping far more Latino drivers than his or her peers. This approach makes no attempt to define an objectively appropriate level of Latino traffic stops (for example, on the basis of the composition of the neighborhood). The data alone do not *prove* that an Officer E is engaged in racial profiling, but do serve as a starting point for a performance review that can determine whether or not his or her activities involve racial or ethnic bias.

The consent decree with the Pittsburgh Police Department directs supervisors "to use the PARS on a quarterly basis to assess allegations of racial bias for patterns or irregularities."[380] On a quarterly basis commanders review the data in the PARS system for an indication of any one of eight indicators of possible bias. These indicators include:

Notation on the Supervisor's Daily Activity Report (SDAR) of any indication of racial or gender bias on the part of a given officer;

A complaint by a supervisor of racial or gender bias against an officer;

A peer complaint of racial or gender bias;

An OMI complaint of racial or gender bias;

Filing of a lawsuit, in which the officer is named, contending racial or gender bias;

Any indication during a normal review of routine police reports (offense reports, arrest reports, search and seizure reports, subject resistance reports, etc.) that an officer shows potential racial or gender bias;

Comments made by an officer indicating racial or gender bias; or

A non-OMI complaint of gender or racial bias.

The court-appointed monitor in Pittsburgh found that supervisors were in fact conducting the required reviews and "during the week of February 4, 2002 identified one officer with at least one of these 'trigger' events."[381] Used on a regular basis, this kind of performance review would not only spot potential racial or ethnic bias very quickly but would also communicate to officers that their performance is being closely monitored. The result would probably be improvement in all types of officer activity and not just with respect to racial or ethnic bias. Whether or not such improvements in fact occur is a question that should be addressed by properly designed evaluations of EI systems.

Intervention

The intervention phase of EI systems consists of counseling or retraining for officers who have been selected. In most EI systems, an officer's immediate supervisor does the counseling. The counseling typically involves a discussion of the officer's performance problems and is intended to lead to some agreement about how the officer might correct these problems. Possible outcomes include advisement from the supervisor, referral to training over particular tactics, or a recommendation that the officer seek professional counseling.

Where remedial training is involved, it is handled by the department training unit. The training might involve a reinstruction on tactics for traffic stops or on handcuffing suspects who resist. In some departments, the counseling session involves other command officers meeting as a committee; some systems use a written performance improvement plan with specific goals.

A few departments—the New Orleans PPEP program, for example—have conducted interventions through a class with a number of officers. Group sessions have a number of problems, however. Difficulty in scheduling a number of officers often results in a delay in holding the classes, and group sessions are not able to focus on the specific problems of individual officers. (The PPEP class does include one component with a private counseling session with each officer.) Finally, bringing together officers who have been identified as having performance problems tends to label them as "bad boys" and give them a sense of solidarity. At least two departments have had counterproductive experiences with "bad boy" classes.[382]

For individualized intervention, some departments provide a specific list of alternative actions that the supervisor can choose from. In the Miami-Dade Police Department, the "Action Alternatives" include referral to the departmental Psychological Services Program or participation in the Stress Abatement Program. Another alternative is a determination that "no problem exists" and that no formal action is necessary.[383] But in many programs, particularly those that were created in the early years without much planning, the interventions are left entirely to the discretion of the immediate supervisor, with no specific list of alternative actions.

Although the intervention stage is the critical component of an EI system, it has received the least attention from EI system experts. This stage is critical to the impact of an EI system because it is where the department delivers the "treatment" designed to improve an officer's performance. Many departments, however, have instituted EI systems with little or no training for supervisors regarding their responsibilities in interventions and no process for holding them accountable for properly conducting interventions.[384]

Inadequate training and supervision for interventions can result in a number of problems. Sergeants may tell officers not to worry about the intervention and that the whole system is just "bullshit." Such a message would completely undermine the purpose of the EI system and breed cynicism about the department's entire accountability effort. Less flagrant but also serious is a situation in which a sergeant sincerely tries but fails to help an officer understand and change his or her performance.

Interventions require certain skills in human relations that all sergeants may not have and for which they are not specifically trained. The sergeant has to firmly but fairly point out the performance problems in an empathetic and nondisciplinary way.[385] Traditional training for supervisors emphasizes the formal and legalistic aspects of discipline: applying the department's discipline code, avoiding grievances and lawsuits, and so forth. In an intervention session a sergeant is expected to coach rather than discipline, for the purpose of helping the officer improve his or her performance. This requires discussing performance deficiencies in a nonthreatening way and suggesting means of improvement.

EI systems have also not given attention to the question of how to hold supervisors accountable for their intervention responsibilities. The matter is complicated by the confidential nature of interventions, which means that there is little documentation of the actual content of intervention sessions.

To date there has been no specific discussion of what action to take when interventions fail to improve an officer's performance and an officer is identified a second (or possibly a third) time by the EI system. Some EI systems include termination as a possible response to identified officers. A convenient alternative is to transfer the officer to an assignment in which he or she will have little contact with the public, and no contact in potentially volatile situations or where use of force might occur (e.g., an internal desk job).

Postintervention Monitoring

Following an intervention, the department monitors the officer's performance for a specified period of time. Postintervention monitoring efforts vary in terms of their formality. Many departments simply keep an officer on the EI system list until a certain period of time has passed (e.g., two quarters) without a significant number of complaints, use of force incidents, or other problem indicators. The New Orleans PPEP program is the most elaborate, with supervisors required to observe an officer at work on a regular basis and to file a formal evaluation every 2 weeks for 6 months.[386] The Pittsburgh PAS system requires sergeants to observe officers (conduct "roll-bys") identified on a regular basis.[387]

GOALS AND POTENTIAL IMPACTS OF EI SYSTEMS

EI systems have a number of different goals and potential impacts. Although they originated as a narrowly focused method of dealing with officers with use of force problems, experts now recognize that they can identify a wide range of performance issues involving individual officers, transform the role of supervisors, and potentially alter the culture of the department as a whole.

Holding Individual Officers Accountable

The original purpose of EI systems was to hold individual officers accountable for documented performance problems and to seek to improve their performance. As is discussed in the next section, the NIJ evaluation found that EI systems have been successful in reducing citizen complaints and officer use of force among officers subject to intervention.[388]

In addition to correcting performance problems, EI system experts increasingly recognize that they also have the capacity to identify and reward good police performance. As in the example of Officer D in Figure 5.2, the data can identify officers with high rates of arrest and traffic stop activity but with few citizen complaints or use of force reports. In Pittsburgh, the commander responsible for the PARS system reports that the system allows them to identify their top performers and their under performers as well as their problem officers. The court-appointed monitor reported that

> Command staff used search and seizure data, generated by PARS, to assess officers' performance, identifying officers who were above average in this category. These officers were noted to be of two types: those who were active in search and seizure processes, but whose reports indicated no problematic behavior, and those whose activities indicated additional training, counseling, or supervision. The command staff selected the first group of officers for potential commendation, and the second group for potential remediation.[389]

In this respect, EI systems have the potential for vastly improving police performance evaluations, substituting fact-based pictures of officer performance for the traditional subjective assessments that involve vague categories such as "works well with people."

Transforming the Role of Supervisors

By their very nature, EI systems transform the role of supervisors, particularly street-level sergeants. This impact (or potential impact) of EI systems was not recognized when the first EI systems were developed, but some experts now believe that it may be one of the most important contributions of EI systems.

EI systems change the role of supervisors in several ways. First, the database gives them systematic data on the performance of the officers under their command. This forces sergeants to become data analysts with an emphasis on identification of patterns of conduct. It is also a radical departure from the traditional reliance on unsystematic impressions, often affected by particularly salient incidents such as a major arrest by an officer. Second, in systems such as the Pittsburgh PARS system, in which sergeants are required to access the system's database on a daily basis, it creates a new standard of intensive supervision. Pittsburgh sergeants are also expected to conduct "roll-bys" of officers who have been identified by the system. Finally, where the system records sergeants' log-ins (as in Pittsburgh), higher-ranking command officers can hold them accountable for their role in the EI system. The court-appointed monitor in Pittsburgh found that sergeants were in fact performing these new duties, with the result that officers with identified problems, and even some supervisors, were being selected for monitoring.[390]

An EI system can also enhance supervision by allowing supervisors to check the past performance histories of officers newly assigned to them. With regular shift changes, sergeants often find themselves responsible for officers about whom they know nothing. As one commander reports, "There is a lot of movement of personnel, so supervisors often do not know the histories of their officers. The EWS report brings them up to speed in a much more timely fashion."[391] Even in a medium-sized department, in which officers' reputations are often known to others, an officer's reputation may not accurately reflect his or her overall performance. The Los Angeles consent decree requires that "when an officer transfers into a new division or area, the Commanding Officer (CO) shall promptly require the watch commander or supervisor to review the transferred officer's TEAMS I [Training Evaluation and Management System] record."[392]

The Vera Institute evaluation of the Pittsburgh consent decree confirmed that the EI system was transforming the role of sergeants. It concluded that the PARS system had "a sweeping change in the duties of the lowest level supervisors."[393] One consequence was that sergeants were spending less time on the street and relatively more time at a desk, staring at a computer screen reviewing data and looking for patterns of conduct.

The consent decree with the Los Angeles Police Department includes specific requirements for the TEAMS II system similar to those in Pittsburgh.

First, it requires that "on a regular basis, supervisors review and analyze all relevant information in TEAMS II about officers under their supervision. . . ."[394] In addition, it requires that "LAPD managers on a regular basis review and analyze relevant information in TEAMS II about subordinate managers and supervisors in their command. . . . ," and that annual reviews of managers' and supervisors' performance will take into account their performance in implementing the provisions of the TEAMS II protocol.

Robin Engel's study of supervisory styles in two large urban departments is particularly illuminating with regard to the role of sergeants. She found that some sergeants do not act as supervisors at all; some see their role as protecting their officers from higher command, whereas some others act as patrol officers themselves, directly engaging in police work.[395] Even among those who do act as supervisors, many are able to act only in a strict disciplinary fashion: enforcing the letter of department rules (that is, playing it "by the book"). If Engel's findings are representative of sergeants across the country, the introduction of an EI system potentially has a significant impact on most American police sergeants, forcing them to change their daily habits and how they define themselves as supervisors.[396]

The San Jose Police Department has taken EI systems to a new level by developing a Supervisor's Intervention Program (SIP) that addresses the performance of supervisors as well as rank-and-file officers. Whenever the team of officers under a supervisor's command receives three or more citizen complaints within a 6-month period, the supervisor is required to meet with his or her chain of command and the head of Internal Affairs. (The San Jose EI system, however, relies only on citizen complaints, which does not represent current best practice.) In the first year of the SIP program, four supervisors met the thresholds and were counseled by the department.[397]

Changing the Organizational Culture

An EI system also has the potential for changing the culture of a department as a whole with respect to accountability. The problem in unprofessional departments is that inappropriate behavior is pervasive and tolerated. To the extent that an EI system involves a serious effort to correct performance problems, it has the potential for altering both the formal and informal norms of the organization.

At this point in the development of EI systems we can only speculate on the impact on police organizations. There have been no studies of this issue, and measuring the impact would be extremely difficult. Even in the best of circumstances, changes resulting from an EI system would be slow and subtle and not evident for some time. Measuring changes in officer use of force would be difficult because implementation of the EI system would probably be accompanied by changes in use of force reporting, making the official data not comparable over time.

EI SYSTEMS AND OTHER POLICE REFORMS

EI systems are closely related to several other important recent innovations in American policing. The following section discusses these parallels.

EI Systems and Problem-Oriented Policing

For all practical purposes, EI systems are a form of problem-oriented policing (POP). As initially formulated by Herman Goldstein, POP is a process through which police departments disaggregate the various aspects of their role and, instead of attempting to address "crime" and "disorder as global categories," should identify particular problems within each category and develop narrowly tailored responses appropriate to each. Goldstein argued that police efforts related to crime and disorder were too vague and nonspecific.[398] The SARA model (scanning, analysis, response, and assessment) that is the framework for problem solving is essentially identical to the EI process. With EI systems, the problem is not graffiti or public drunkenness but those officers with performance problems. Scanning involves a review of the performance data in the EI system database. The analysis involves the identification and selection of officers who need intervention. The response is the actual intervention with the selected officers. Finally, the assessment phase is the postintervention review of those officers' performance.

EI Systems and COMPSTAT

EI systems are also similar to the purpose and processes of COMPSTAT programs. One of the most important innovations in policing in recent years, COMPSTAT involves the collection and analysis of systematic data on crime and disorder for the purpose of identifying patterns and then developing appropriate responses in terms of police strategies and tactics. At the same time, it is designed to heighten the accountability of precinct commanders by pinpointing

---

Scanning

Review of performance data in the EI System database
Identification of officers with apparent performance problems

Analysis

Performance evaluation of officers identified in the scanning phase
Selection of officers for intervention

Response

Intervention: counseling, retraining, etc.

Assessment

Postintervention review of performance of officers subject to intervention

---

Figure 5.3   Applying the SARA model to EI systems.

---

the problems under their command, requiring them to develop the responses, and regularly assessing the effectiveness of their actions.[399]

In a similar way, an EI system provides systematic data on officer performance and requires their supervisors to take the appropriate corrective action. Whereas COMPSTAT looks outward, defining the problem as crime and disorder, EI systems look inward, defining the problem as officers with recurring performance problems. The similarity between EI systems and COMPSTAT was recognized in the VERA evaluation of the Pittsburgh PARS system (its EI system). The evaluation described one key element as modeled after COMPSTAT. The Quarterly COMPSTAR meetings in Pittsburgh involve a review of officers who have been identified by the PARS system. Area commanders make presentations about any officers under their command who have been identified and conclude with a recommendation regarding formal intervention. After a discussion, the chief of police makes a final decision on what course of action to take.[400]

EI Systems and Risk Management

An EI system is essentially a risk management system. The concept of risk management, which is well developed in the private sector and in medicine, is a process for reducing an organization's exposure to financial loss due to litigation. It involves data collection and analysis to identify the areas of financial risk, to examine the underlying causes of these problems, and then to correct those problems through improved policies, procedures, and training.[401]

Risk management has not been widely adopted in policing, despite the fact that cities and counties experience significant civil litigation costs because of fatal shootings, excessive physical force, and high-speed pursuits. An EI system database can serve a risk management program, readily identifying patterns of individual officers and situations that represent actual or potential risks. In fact, the expanded EI system in the Cincinnati Police Department, as mandated by a memorandum of understanding with the Department of Justice, is called the Risk Management System. Although an EI system focuses on the performance of individual officers, additional risk management procedures are necessary to ensure that the department takes the necessary steps to revise policies and training to prevent future problems.

THE EFFECTIVENESS OF EI SYSTEMS

The available evidence suggests that some EI systems have been successful in achieving their goals of reducing officer misconduct. At the same time, there are some EI systems that have not been successfully implemented.

The NIJ Evaluation of Three EI Systems

An NIJ study of EI systems in three police departments found significant reductions in use of force and citizen complaints among officers following EI intervention. First, the study found that in both Minneapolis and Miami-Dade, the EI systems successfully identified officers whose performance records were significantly worse than those of peer officers who had been hired in the same years. In Miami-Dade, officers selected by the department's early intervention system (EIS) averaged twice as many use of force reports per year as non-EIS officers and were almost three times as likely to have ever been suspended. In Minneapolis, officers selected by the early warning (EW) system averaged twice as many complaints as their non-EW peers and were three times as likely to have been suspended.[402]

The successful identification of officers whose disciplinary records are significantly worse than their peers is an important finding. There has been much concern among police officials that an EI system may identify active and productive officers (who, because of their arrest activity, are more likely to use force and receive complaints than less-active officers), and as a consequence deter them from active police work. The Vera Institute study of the implementation of the consent decree in Pittsburgh, which included the introduction of an EI system, found no evidence of reduced police activity, or what it termed "depolicing."[403]

The NIJ study found that in both Miami-Dade and Minneapolis, the EI systems had a positive impact on officers' performance. In Minneapolis, the average number of complaints per year received by EI officers fell from 1.9 prior to intervention to 0.65 following intervention. In Miami, 27 of the 28 EIS officers had at least one use of force report; following intervention only, 14 had a use of force report. Twenty of the 28 officers had four or more complaints prior to intervention, whereas only 9 of the 28 had four or more complaints following intervention.

The nature of the New Orleans Professional Policing Enhancement Program (PPEP) program permitted direct observation of the training class for problem officers and an analysis of their assessments of the class. Observation of the class found that officers responded in very different ways to different units and teaching methodologies. Most of the officers expressed open hostility to the PPEP program at some point, referring to it as "politeness school," or "bad boys class." They were visibly disengaged from units that involved lectures, and in particular lectures about general topics such as the role of the police, stress management, and the problem of substance abuse in American society.[404]

At the same time, however, the officers were actively engaged in the unit of the course that asked them to critique case studies drawn from the department's files. They appeared to be very knowledgeable and professional police officers, fully aware of proper procedure in difficult situations and proud of their own expertise about how to handle such situations. This engagement and sense of professionalism was exhibited by the same officers who had earlier voiced extreme hostility to the PPEP program.

Officers in the PPEP classes were given an opportunity to complete a written evaluation of the class. A total of 26 evaluations were available for analysis. The officers gave the PPEP classes extremely positively ratings. On a scale of 1–10, they gave it an average rating of 7.0. When the four officers who gave it the lowest possible rating (1) are omitted, the average rating rises to 8.1. These responses are extremely significant in light of the fact that the officers in the observed class openly disparaged the PPEP program. Narrative comments on the evaluation forms also included highly favorable assessments of the class. All of the officers made at least one positive comment about the class. Comments included favorable references to the verbal judo and complaint reduction component and the stress reduction component. About three-quarters of the officers also made negative comments, but all were related to the PPEP program in general or the department and its leadership and not about the content of the PPEP class itself.

In short, the evidence from the New Orleans program strongly suggests that, despite their public posture of hostility to such efforts, officers appreciate efforts to help them improve their performance. This finding is consistent with the lack of active opposition to EI systems reported by police managers in the PERF survey discussed below.

The Los Angeles Sheriff's Department's PPI System

The PPI system in the Los Angeles Sheriff's Department has been widely regarded as possibly the best EI system in the country. LASD representatives have been invited to give presentations about the system at national conferences sponsored by the U.S. Department of Justice and other organizations. Early reports by the Special Counsel to the Los Angeles Sheriff's Department indicate positive outcomes for the PPI system. Among officers subject to performance review (PR) the rate of officer-involved shootings dropped from an average of .50 per month before PR to zero following PR. Use of force incidents dropped from an average of 7.11 per month before PR to .98 per month after PR. Finally, personnel complaints dropped from an average of 3.86 to .74 in the same time periods.[405] These are very substantial improvements in police performance with enormous implications for police–community relations.

In addition, 11% of all officers placed on performance review left the department within a short period of time. This included ten retirements, six discharges, one resignation, and one departure for unknown reasons.[406] It is not unreasonable to conclude that the PPI influenced these decisions, sending a strong message to the officers that their performance did not meet department expectations. In short, it had a deterrent effect on misconduct. These departures not only saved the department the immediate expense of conducting a full 2-year performance review but also undoubtedly reduced the risk of future serious misconduct by one or more of these officers.

A 2003 report by the LASD Special Counsel, however, identified some serious management problems with the PPI system that call into question its reputation and current effectiveness. That report is discussed below.

The Experiences and Perceptions of Police Managers

In 2002, the University of Nebraska at Omaha in collaboration with the Police Executive Research Forum (PERF) conducted a national survey of PERF members regarding their perceptions and experiences with EI systems. A total of 135 managers responded to the survey, about 40% of whom had experience with an EI system. The responses from more than 50 police managers with direct experience with EI systems represent the most comprehensive assessment of how EI systems are operating.[407] The findings were both very positive and surprising in certain important respects.

Impact on the Quality of Police Service

Police managers overwhelmingly report that their EI system has had some positive impact on the quality of on-the-street police service (Table 5.2). Almost half (49%) report a positive impact and 28% report a mixed impact, for a combined total of 77%. Perhaps even more important, no manager reported a negative impact on the quality of police service. Some skeptics have suggested that the heightened scrutiny of performance in EI systems might cause officers to reduce their level of activity to avoid potential citizen complaints or use of force incidents—a phenomenon labeled "de-policing."[408] But no manager in the survey reported that the EI system caused officers to back off and reduce their activity level.

Impact on the Nature and Quality of Supervision

Managers in the survey report that their EI system has enhanced supervision in their department. Many of them offered specific examples. One explained that "Sergeants have been able to evaluate the strengths and weaknesses of their squad even before meeting with them; they are able to develop a proactive strategy to address personnel and leadership issues." Another explained that it encouraged proactive supervision: "Supervisors pay more attention to what is going on, document all activities more thoroughly, and talk

Table 5.2   Impact of the EI System on the Quality of Police Service
                  Perceptions of Police Managers

Perceived Impact	Percentage
Positive	49%
Mixed	28%
Negative	0%
No impact	23%

Note. From Early Intervention Systems for Law Enforcement Agencies: A Planning and Management Guide (p. 75), by Samuel Walker, Washington, DC: U.S. Department of Justice, 2003.

with officers when things might look like a problem." A third added that the "Program provides a way for the department to provide nondisciplinary direction and training before the officer becomes a liability to citizens, the department, and him/herself." One manager in the survey confirmed the capability of an EI system to help supervisors familiarize themselves with newly assigned officers, explaining that "There is a lot of movement of personnel, so supervisors often do not know the histories of their officers. The EWS report brings them up to speed in a much more timely fashion."

Several managers commented that the database facilitated a new style of supervision. One, for example, commented that the EI system is "a useful tool to involve supervisors and lieutenants in a nontraditional model of problem solving. It has served to enhance their management skills and help round out their people-interaction skills." The comments by other managers help to amplify this point, explaining that the database allows supervisors to talk to officers about specific performance problems and to avoid generalities or statements that might otherwise be dismissed by an officer as unfounded rumor. Finally, by providing systematic data on all officers, the EI system enhances fairness. One manager commented that it "levels the playing field. No one can be accused of playing favorites."

Reactions of Rank-and-File Officers

The most surprising finding is the managers' perception of rank-and-file officers' response to their EI systems. Contrary to the expectations of many people, the managers do not report significant opposition from rank-and-file officers or from police unions.[409] The negative reactions that do exist primarily involve an initial fear of the unknown and lack of knowledge about the EI system. One manager commented that "The system was met with much cynicism and distrust when first introduced. [But] experience and education has reduced those attitudes." Another observed that "Most [officers] no longer believe this system is out to 'get' them, but rather to assist them." Only 16% of managers reported any serious opposition from police unions, and there is no record (either from the survey or any other source) of a police union succeeding in blocking the operation of an EI system once it is in place (although in some instances, the union has insisted that development of an EI system is a bargaining issue).[410]

A comment by one manager suggests one possible explanation about the lack of strong rank-and-file resistance. This manager observed that "Nothing hurts morale like no action being taken against problem officers." An EI system shows "that problem people can be dealt with." Insofar as an EI system does identify officers with performance problems and leads to some corrective action, other officers will develop a more positive attitude about the department and its commitment to professional standards. And as mentioned earlier, part of the folklore of the police subculture is that peer officers do in fact know who the problem officers are. A contributing factor to officer cynicism is the perception that bad officers are not punished and good performance not rewarded. In this respect, a properly functioning EI system may contribute significantly to the morale of the better officers.

Problems With Planning and Implementation

Police managers report that the major problems with EI systems involve failures in planning and implementation, but not with the basic concept of an EI system. One manager commented that "While the system was crafted appropriately, it was not explained to the officers or the first line supervisors to the extent necessary to make it an understandable and viable system." Along the same lines, another manager complained that department leaders "did not explain the purpose of the program well." Finally, a number of managers felt that the department did not follow through in terms of fully using the potential of their EI systems. According to one, "The program does not hold a prominent place in the organization. As a result, no results are provided to commanders as to its effectiveness or benefits." The following section examines some of problems in follow-through that have occurred in some EI systems.

In the end, the significant aspect of the PERF survey is the overwhelmingly positive perceptions of police managers toward EI systems. There is no evidence that managers question the basic concept of EI systems or have found that it is dysfunctional when operating as planned.

PROBLEMS IN IMPLEMENTING AND MANAGING EI SYSTEMS

Although EI systems have enormous potential for correcting officer performance problems and developing a culture of accountability in a police department, there is also disturbing evidence that some systems have not been effectively implemented and used. In some instances, departments did not implement the EI systems they claimed to have. In other cases, the EI systems were woefully inadequate. And in at least one other notable case, a very good system suffered from administrative neglect.

Three Cases of Failure

Pennsylvania State Police

In its 2002 *Annual Report,* the Pennsylvania State Police described a functioning early intervention system. But when a sexual abuse scandal erupted in 2003, media inquiries revealed that the EI system was still only in the process of being developed.[411]

Miami Police Department

The Miami (FL) Police Department established one of the first EI systems in the early 1980s (see Chapter Three), and this system has maintained continuous existence through to the present. A corruption and brutality scandal that erupted in 2000, however, raised serious questions about whether the system functioned at all. Eleven officers were indicted on federal criminal charges, and several were eventually convicted. Some of the officers had substantial histories of misconduct that should have been detected by an EI system. The U.S. Department of Justice investigated the department and found many serious deficiencies related to accountability. With respect to its early intervention system, the Department of Justice concluded that "It is clear that the MPD recognizes that its EWS needs to be improved, and we understand that significant changes to the EWS are contemplated."[412] Among other things, the Department of Justice found a pattern of under-reporting of use of force incidents and a practice of discouraging citizens from filing complaints. In short, the department was not collecting all of the critical performance data that are necessary to make an EI system an effective accountability mechanism.

Albuquerque Police Department

An evaluation of the Albuquerque Police Department by the Police Assessment Resource Center (PARC) found a number of problems with its EI system. The system was managed by a single part-time volunteer. There was already a backlog in entering use of force reports in this relatively new system. In addition, the system functioned through a single computer located in Internal Affairs. As a result, "Access to the information in the early warning system is not readily available to APD commanders and managers, except through requests for reports to the volunteer." Moreover, no one in the police department had been trained either to assist or take over from the volunteer in the event of his unavailability or departure. In operation, it was not clear whether commanders met with all officers identified by the EI system, and they did not document what actions they took with regard to those officers.[413]

The Saga of TEAMS II in The Los Angeles Police Department

The saga of the TEAMS (Training Evaluation and Management System) system in the Los Angeles Police Department (LAPD) is an especially disturbing story of implementation failure occurring even while a department is under intensive public scrutiny.[414] Following the beating of Rodney King on March 3, 1991, the Christopher Commission identified a small group of 44 officers that had particularly serious performance records and recommended the creation of an EI system to address this problem.[415] Implementation of the TEAMS system over the next decade is a story of inaction. In a 1996 report to the Los Angeles Police Commission, Merrick Bobb reported that the LAPD did not then have an operational EI system as recommended by the Christopher Commission and that the TEAMS system was "weak and inadequate."[416]

The Los Angeles Police Commission then sought and received a $175,000 federal grant to facilitate the implementation of an expanded TEAMS (to be called TEAMS II). But in the wake of the Rampart scandal that erupted in 1999, a subsequent Police Commission report found that TEAMS II was not currently operational. In fact, the city had not even drawn on the federal grant awarded to implement TEAMS II. Moreover, LAPD Police Chief Bernard Parks had previously announced with much fanfare that almost all of the Christopher Commission recommendations had been implemented. The Rampart scandal exposed the emptiness of that claim and prompted a Department of Justice pattern or practice suit against the LAPD. The subsequent consent decree in 2001 directed the department to complete the implementation of TEAMS II .[417]

The saga of TEAMS I and II is particularly disturbing with respect to the possibilities of achieving meaningful accountability measures in a large police department. The ominous implications of the story, in fact, apply not just to EI systems but to the full range of related policies and procedures, such as use of force reporting systems, citizen complaint procedures, the review of misconduct reports, and so on. In addition, the court-appointed monitor reported in late 2003 that "The ability to conform to the original timeline for completion of the TEAMS II project has, from the inception of the Monitorship, been very much in question." It did note, somewhat more optimistically, that the city was currently "moving forward in as expeditious a manner as possible" to implement TEAMS II.[418]

In short, 13 years after the initial Christopher Commission recommendation, the LAPD did not have a fully operational EI system. (This book was completed in mid-2004.) It is especially alarming that this failure could occur in the context of enormous public scrutiny. Perhaps the failure to implement TEAMS or TEAMS II can be attributed to the well-known LAPD culture of resistance to external criticism and direction, but it may also have broader application to large departments.

The Case of the Los Angeles Sheriff's Department's PPI System

The history of the Personnel Performance Index (PPI) in the Los Angeles Sheriff's Department (LASD) is a disturbing tale of inadequate maintenance of an established system. The PPI has been widely regarded as the most sophisticated EI system in the country. Merrick Bobb, Special Counsel to the LASD and the most knowledgeable outside expert on the department, described it as "without question, the most carefully constructed and powerful management tool for control of police misconduct currently available in the United States."[419] LASD officials have been invited to speak at conferences around the country, and the PPI has been presented as a model program that other departments should emulate. In addition to a comprehensive set of performance indicators, the PPI system (as described earlier in this chapter) operates in a two-stage discretionary manner that includes a full-scale performance review of officers initially identified by the system followed by a decision about whether to refer particular officers for intervention.

In early 2003, however, Bobb published a blistering report identifying serious problems with the management of the PPI. First, crucial performance data were not being entered into the system, and reports that were being sent to it "are often sloppy and error-ridden." Many of the reports were initially rejected because they contained errors. Bobb estimated the error rate at 50% for one category of reports. With respect to citizen complaints, for example, Bobb found uncertainty within the department "about what constitutes a citizen's complaint," with the result that "there are inconsistencies in what gets reported." This problem highlights the fact that an EI system is only as good as a department's reporting systems related to use of force, citizen complaints, and other performance issues.[420]

Second, there were "long delays in entering information into the computer database." The staff were taking about five and a half months to enter citizen complaints into the PPI. During the previous summer, for example, the Century, Lennox, Compton, and East Los Angeles stations and the Twin Towers jail turned in more than 350 force reports "that were long overdue," some of which dated back to 1996. One of the major causes of these delays was understaffing in the critical administrative positions. Bobb found ten unfilled positions in the Central Custodian Records Unit, which processes the data for the PPI.

Third, and even more alarming, some area commanders were not aware of the PPI's capabilities. Bobb found one station commander who assigned a civilian computer programmer "to create a database to provide reports on citizen's complaints and deputies' use of force." The programmer spent 1,000 hours developing a database that duplicated part of the PPI. Another captain complained that the PPI was of limited use because it did not display details regarding citizen's complaints. This captain had not been instructed how to use the PPI to bring up those details, including the reports from the underlying investigation.[421]

In a stinging conclusion, Bobb commented that "The LASD currently treats the PPI like a collectible automobile: It is put on display from time to time to demonstrate to the outside world that the LASD has the Rolls Royce of risk management software and procedures. And indeed it is the Rolls. But when the odometer is checked, it is apparent that it has hardly ever been taken out of the garage." He concluded that despite the PPI's great potential, "Today, its resources are largely untapped."[422]

The Case of the Pittsburgh PARS System

The PARS system in the Pittsburgh Police Bureau, on the other hand, represents a substantial success story. Implementation of the system has been subject to two separate monitoring and evaluation procedures. The consent decree settling the federal pattern or practice suit required the appointment of an independent monitor to oversee implementation of the decree. The monitor found the Pittsburgh Police Bureau to be in compliance with the decree, including the provisions related to the PARS system. In 2002, the monitor reported that "PARS is now completely operational, and was used by the PPB command staff to conduct its last quarterly analysis during the month of February 2002."[423] The federal court dissolved the consent decree over the department in mid-2002. The Vera Institute evaluation of the impact of the consent decree, meanwhile, found that the PARS system had been successfully implemented, was being used, and, as noted earlier in this chapter, was having a significant impact on sergeants' activities.[424]

CONCLUSION

Early intervention systems are an important new innovation in police management. Experts view EI systems as the centerpiece of the new police accountability. Properly implemented, EI systems have the potential for identifying officers with recurring performance problems and correcting their performance. At the same time, they have the potential for dramatically transforming the role of sergeants in a positive direction. Finally, there is the possibility, as yet unconfirmed by independent research, that they can transform the organizational culture of a police department and help to instill new standards of accountability.

Although they have great potential, the experience to date clearly indicates that EI systems are extremely complex administrative mechanisms and require an enormous investment in both money and management attention if they are to function effectively. Even in this early stage of the history of EI systems, there are examples of departments that failed to implement an EI system, implemented one in an inadequate way, or failed to maintain a system that was well established.

SIX

THE AUDITOR MODEL
OF CITIZEN OVERSIGHT

————•◆•————

E xternal citizen oversight of the police has been one of the principal demands of civil rights activists since the 1960s. This demand has usually been for a civilian review board, an independent agency that would investigate citizen complaints against the police. In the 1990s, a new form of citizen oversight emerged: the police auditor.[425] This chapter argues that the police auditor is more likely to be an effective form of citizen oversight than the traditional civilian review board, and even more important, the police auditor has the potential for ensuring the long-term success of accountability efforts.[426]

The crucial aspect of the police auditor concept is its focus on organizational change. This distinguishes it from civilian review boards, which investigate individual complaints. Instead of focusing narrowly on the culpability of officers in particular misconduct incidents, police auditors focus on organizational problems that underlie such incidents. Mike Gennaco, head of the Office of Independent Review (OIR) in the Los Angeles Sheriff's Department, explains the strategy guiding his office:

> To change behavior effectively, an oversight body *must look beyond the particular cases of misconduct to systemic issues implicating policy and training* [emphasis added]. For example, ambiguities in policy or lax enforcement of an existing policy can prevent LASD from imposing discipline. Alternately, insufficient training on a policy, procedure, or legal issue can lead to inadvertent violations. Deputies must know the standards they are held to and LASD must exhibit even-handed enforcement of policy violations.

Accordingly, OIR endeavors to use individual cases to identify ambiguities in policy, laxity in enforcement, and deficiencies in training. Whenever policies and practices can be reformed to eliminate potential civil rights violations and future liability, it will directly benefit the people of Los Angeles County.[427]

Police auditors have two special capabilities that enhance their ability to promote organizational change. First, as full-time government officials they have the authority to probe deeply into departmental policies and procedures with an eye toward correcting them and reducing future misconduct. Second, as permanent agencies they can follow up on issues and determine whether or not prior recommendations for change have been implemented. The capacity for sustained follow-up addresses not only the historic limitation of blue-ribbon commissions,[428] but also one of the most serious problems in police accountability: how to ensure the implementation of recommended reforms and sustain reform over the long-term.

This chapter gives special attention to several police auditors, particularly the Special Counsel and the Office of Independent Review in the Los Angeles Sheriff's Department and the San Jose Independent Police Auditor. These agencies have the most creditable records of activity to date and demonstrate the potential of a police auditor's office. At the same time, however, some police auditors have been notable failures. The reasons for their failure are discussed in a section on the limitations of the police auditor concept.

POLICE AUDITORS IN THE UNITED STATES

There are 12 police auditors in the United States, covering 11 law enforcement agencies (the Los Angeles Sheriff's Department has two separate auditors). This represents about ten percent of all citizen oversight agencies. The precise number of police auditors is somewhat ambiguous because there is no clear or universally agreed-on definition of the concept. As is also the case with civilian review boards, the structure and function of existing auditors vary considerably. Also complicating the picture is the fact that some police auditor functions are also performed by a few civilian review boards. Despite these complexities, the core auditing function is clear: the capacity to review police department policies and procedures.

A glance at Figure 6.1 clearly indicates that police auditors are clustered in the western half of the United States. This pattern reflects the distribution of

Austin (Texas) Police Monitor

Boise (Idaho) Community Ombudsman

Los Angeles County (California) Sheriff's Department, Special Counsel

Los Angeles County (California) Sheriff's Department, Office of
Independent Review (OIR)

Nashville (Tennessee) Metropolitan Police Department, Office of
Professional Accountability

Omaha (Nebraska) Public Safety Auditor

Philadelphia (Pennsylvania) Integrity and Accountability Office (IAO)

Portland (Oregon) Independent Police Review Division

Sacramento (California) Office of Police Accountability (OPA)

San Jose (California) Independent Police Auditor (IPA)

Seattle (Washington) Police Department, Office of Professional
Accountability (OPA)

Tucson (Arizona) Independent Police Auditor (IPA)

Figure 6.1   Police auditors in the United States

all citizen oversight agencies in this country. For a variety of reasons that we
can only speculate about, cities and counties in western states have been more
open to external oversight of the police, whereas those in the northeast have
been most resistant.[429]

Police auditors originated as a political compromise—initially in San Jose
and Seattle in 1993. In both communities, local activists pressed for the creation of a civilian review board, but in the face of strong opposition from local
police unions the city councils created police auditors as a compromise
that was less threatening to the police. Because of these origins, civil rights
activists have always regarded the concept of police auditors with some
suspicion, particularly because they do not investigate individual citizen complaints.[430] This chapter argues that this criticism of police auditors fails to recognize the special contributions they can make with regard to achieving greater
police accountability.

Authority and Structure

The existing police auditors differ in terms of the formal basis of their authority and their organizational structure.[431] Most are established by ordinance, some operate under renewable contracts, and one was created under the terms of a consent decree. The San Jose Independent Police Auditor (IPA) is a separate municipal agency, and because it is now specified in the city charter (as opposed to being based on a simple ordinance) it is probably the most independent of all the auditors, with considerable protection against abolition or interference.[432] In practice, this means that auditor Teresa Guerrero-Daley has been able to issue reports identifying serious problems in the San Jose Police Department (SJPD) without fear of meaningful reprisal from either the police department or the mayor's office. The Boise Ombudsman and the Omaha Public Safety Auditor are both established by ordinance, but, not being grounded in city charter as is the San Jose IPA, they do not enjoy the same level of political independence.[433]

Both the Special Counsel (SC) to the Los Angeles Sheriff's Department and the Office of Independent Review (OIR) for the LASD operate under contracts, with the County Board of Supervisors and the Sheriff's Department, respectively.[434] They are the least independent police auditors because they can be terminated simply by not renewing their contracts (although, ironically, they have two of the best records of accomplishment to date; see below). The Philadelphia Integrity and Accountability Office (IAO) was created as part of the settlement of a civil suit against the police department alleging excessive use of force.[435] Whereas the office is mandated by the court, the director is formally an employee of the police department and thereby occupies an awkward status as both an "outsider" and an "insider." The Seattle Office of Professional Accountability (OPA) is also an outsider–insider hybrid. It was created by an ordinance requiring the OPA Director to be a nonsworn person appointed by the police chief to a 3-year term, with the rank of assistant chief. The director is responsible for directing the police department's internal affairs office.[436] Nashville also created a civilian-staffed Office of Professional Accountability (OPA) in 2000.

Functions and Activities

Police auditors perform five basic functions. As already indicated, the investigation and review of citizen complaints is not one of those primary functions, although some auditors do become involved in particular cases. This section will identify and describe those functions briefly. Later sections that describe particular police auditor offices provide a far better sense of what police auditors can accomplish.

1. Auditing the Complaint Process

Police auditors audit or monitor the police department's citizen complaint process. This includes reviewing the procedures by which the department publicizes the complaint process, receives complaints, and records and classifies complaints, as well as reviewing patterns of complaints. In one of its first actions, for example, the San Jose IPA found that the police department was improperly classifying many complaints, with the result that the number of serious complaints was artificially low. A policy recommendation resulted in a new system for classifying and recording citizen complaints (and a resulting increase in the official number of serious complaints).[437] In this respect, a police auditor functions as an external monitor ensuring that a police department's complaint process meets the standards described in Chapter Four of this book.

2. Auditing Police Operations

Police auditors also investigate basic police operations, and it is in this capacity that they have the greatest potential for enhancing police accountability. The Philadelphia Integrity and Accountability Office (IAO), for example, examined narcotics enforcement by the Philadelphia Police Department and found a number of serious deficiencies. The department did not collect "critical information that would enable it to monitor the integrity and effectiveness of its narcotics enforcement activities." Even when it did identify problems, it failed to take appropriate action to correct the problems. The personnel selection process for the Narcotics Bureau was so inadequate that officers "ill suited" for this kind of work were assigned to the Bureau. Training for narcotics officers was "inconsistent and sporadic." Disciplinary practices in the Philadelphia Police Department were "lax and inconsistent." Supervisors tolerated a variety of practices that allowed or even encouraged officers to "cut constitutional corners." In short, the IAO report painted a picture of a critical police unit that was functioning without meaningful standards. The role of the IAO is to examine such issues with an outsider's perspective and bring the problems to public attention.[438]

The Philadelphia IAO report highlights the crucial difference between the police auditor and civilian review board approach to reducing police misconduct. Even if an aggressive review board succeeded in obtaining discipline for many individual officers in the Narcotics Bureau, the underlying mismanagement of the unit, with its lack of concern about accountability, would continue, with the result that officer misconduct would continue.

3. Policy Review

Audits of the complaint process and of police operations lead to recommendations for changes in police policy. Policy recommendations can also arise from individual citizen complaints. This process, known as *policy review,* is potentially the most important function that any citizen oversight agency can perform because it is directed toward organizational reforms that will prevent future misconduct.[439] Policy recommendations are sent to the police chief or sheriff. In some jurisdictions the chief or sheriff is required to respond in writing. Policy recommendations are not binding. In its first 10 years of operation, however, the San Jose IPA made a total of 81 recommendations, and the San Jose police department rejected only seven (several were still pending as this is written). This acceptance rate indicates a high level of trust and cooperation between the IPA and the police department and clearly suggests that the IPA recommendations were not unreasonable.[440]

Because the chief is not required to accept an auditor's recommendation, critics of the auditor model argue that it is essentially "toothless." The policy review process has other important values, however. The process of investigating problems, of making public policy recommendations, and requiring a public response from the department represents a professional and civilized dialogue over police issues. This is an alternative to the traditional process in which charges and countercharges about police misconduct are exchanged in the media, with much posturing on both sides and with the public receiving little if any useful information about the underlying issue. Merrick Bobb, the LASD Special Counsel, explains the additional value of the private dialogue that occurs with top police management:

> The dialogue between us [the LASD Special Counsel] and the Department
> has been primarily an exchange of views about cause and effect—we tend to
> emphasize the degree of management control and the LASD emphasizes the
> danger of the environment in which Century deputies operate. At least, that

is true of the *public* dialogue between us. Our private dialogue has been different and more earnest in tone, with the focus mainly on efforts by management at Century and in the Department as a whole to reduce force used by deputies at the station.[441]

4. Community Outreach

One of the long-standing criticisms of American police departments has been that they are closed bureaucracies that provide little information about their activities and are hostile toward critics and citizen complainants in particular. The community outreach function of police auditors is designed to overcome this problem. Outreach activities include meeting with community groups, particularly those with historic problems with the police; providing information about how to file a complaint; and listening to community concerns about the police department. Because the auditor's office is independent of the police department, it has greater credibility with community groups than do representatives of the police department.

The San Jose IPA and the Austin, Texas, Police Monitor have maintained active programs of community outreach. Notably, the San Jose IPA published a very readable booklet on police relations with juveniles, *A Student's Guide to Police Practices,* with detailed advice to young people on their responsibilities in dealing with police officers.[442] The Seattle OPA sponsored a similar booklet, *Respect: Voices and Choices,* published in eight languages other than English.[443]

5. Creating Transparency

Police auditors also provide greater openness—or transparency—for the police departments they serve. The historic closed and secretive nature of American police organizations has been a combination of two factors: the inherent nature of all bureaucracies to be self-protective in the face of outside inquiry, and the unique tradition of the police subculture. Community policing, it should be noted, is designed to overcome this problem. Until relatively recently, it has often been difficult for outsiders, including even responsible public officials, to obtain information on police operations. Police auditors change this by providing periodic public reports on both their own activities and important aspects of the law enforcement agency they are responsible for.

The San Jose IPA is required by ordinance to file quarterly public reports that include a statistical analysis of citizen complaints, an analysis of complaint trends and patterns, and recommendations for changes in policies and procedures.[444] The Boise Community Ombudsman, meanwhile, has issued a series of detailed reports on controversial incidents (e.g., shootings, sexual assault investigations) that separate the facts from the rumors and misunderstandings and include recommendations for changes in police operations where appropriate.[445]

A comparison of annual reports from different agencies illustrates the role of police auditors with regard to transparency. The 18 Semiannual reports (as of mid-2004) issued by the Special Counsel to the Los Angeles Sheriff's Department provide a wealth of information on virtually every aspect of departmental operations.[446] In a similar way, the reports of the San Jose Independent Police Auditor provide considerable detail about the San Jose Police Department. The reports of the New York City Civilian Complaint Review Board (CCRB), by comparison, contain a wealth of statistics on complaints, but little about the policies and procedures of the police department that underlie recurring problems that are of great concern to the community.[447] The reports of the Philadelphia Police Advisory Commission (PAC), a traditional civilian review board, meanwhile, contain very little about the complaint process or the police department. The annual reports of the Kansas City Office of Citizen Complaints (OCC), another traditional review board, tell us virtually nothing about either the Kansas City Police Department or what the OCC did in the previous year.[448]

Case Study: A New Standard in Openness on Police Discipline

One of the notable aspects of police discipline is that it is a completely closed world, about which outsiders are able to learn almost nothing. Aside from a small number of departments that have adopted a discipline matrix or schedule of discipline, it is impossible to know what the likely punishment would be for a particular act of misconduct (e.g., an excessive use of force that did not result in injury). And in most departments the discipline imposed in particular cases is not public information. In some departments the union contract specifically forbids the department from releasing such information. Finally, no one has an idea of what the "going rate" (the normal punishment) is for particular acts of misconduct. This situation contrasts sharply with the criminal courts, where the state criminal code specifies the possible punishment, individual sentences are matters of public record, and research can readily determine what the going rate is for particular crimes (and the sentencing practices of particular judges, and so on).

The lack of openness or transparency is the source of much community discontent because community leaders may believe that officers are not punished, or not punished appropriately, for their misdeeds. Some oversight agencies have changed that practice and made public the disciplinary actions of the department they oversee. The most notable example of this practice is the Office of Internal Review (OIR) in the Los Angeles Sheriff's Department. Several times a year the OIR publishes (and posts on the Web) a review of disciplinary actions.[449] The reports include the following:

- a summary of the incident and the alleged misconduct
- a summary of the OIR recommendation in the case
- the result of the LASD's internal investigation
- the disciplinary actions taken
- the subsequent history of the case
- a summary of whether any criminal or civil legal action arose from the case

Auditors, Blue-Ribbon Commissions, Monitors, and Review Boards

The role of the police auditor in the larger structure of the new police accountability is clarified if we compare it with other external oversight mechanisms.

Blue-Ribbon Commissions

Police auditors represent a very significant improvement over the traditional blue-ribbon commission.[450] As explained in Chapter Two, blue-ribbon commissions have neither the capacity to implement their recommendations nor to monitor implementation by the police department.[451] Police auditors, on the other hand, are permanent agencies and have the capacity to conduct follow-up investigations on prior reports, monitor the progress of implementation, and not only identify failures but pinpoint the underlying reasons. The importance of this follow-up capacity is particularly well illustrated by the LASD Special Counsel, which has identified several cases of reforms that were allowed to deteriorate (see the discussion below.)

Police Monitors

Police auditors should not be confused with court-appointed police monitors, which are related to settlements of pattern or practice suits against police
departments. A monitor is an agent of the court and its investigating authority
is limited to the specific terms of the consent decree. Moreover, its life span is
limited to the term of the decree, typically 5 years.[452] Police auditors, on the
other hand, are permanent agencies, free to continue their work indefinitely,
and are not limited in terms of the issues they can investigate. The discussion
of the LASD Special Counsel and the San Jose IPA later in this chapter examines the wide range of issues they have investigated in their respective law
enforcement agencies.

Civilian Review Boards

As previously discussed in Chapters Two and Four and at the outset of this
chapter, police auditors are different than civilian review boards in that their
primary mission does not include the receipt, investigation, and review of
citizen complaints against the police. Review boards do not have the power
to impose discipline themselves, but are limited to forwarding a disposition
(sustained, not sustained, unfounded, exonerated) to the police chief executive.[453]

The differences in mission reflect different visions of police reform and
how to achieve police accountability. Civilian review boards reflect a *criminal
process model* of citizen oversight, directed toward investigating citizen complaints for the purpose of punishing guilty officers. The underlying assumption
is that the investigation of complaints by persons who are not themselves
sworn officers will produce more thorough and independent investigations,
with the result that more complaints will be sustained and more officers disciplined. More discipline, in turn, is expected to deter future officer misconduct
on the part of both the officers disciplined (specific deterrence) and other officers (general deterrence).[454]

To date, however, there is no evidence that civilian review boards are
effective in achieving their stated goals. They do not sustain a significantly
higher rate of complaints than police internal affairs units. Both internal and
external complaint procedures sustain about 10–13% of all complaints.[455]
There are readily understandable reasons for this. As a number of experts have
argued (including this author), citizen complaints against police officers are
inherently difficult to sustain. Often there are no independent witnesses to the

incident, and except for the few very egregious physical abuse cases, there is rarely any forensic evidence such as medical records. Officers can always claim that force was necessary to overcome serious resistance and threats to his or her safety by the citizen. Most complaints end up as swearing contests between citizen and officer. These constraints affect external as well as internal citizen complaint investigative procedures and severely limit their capacity to sustain individual complaints.[456]

Another limitation of the civilian review board focus on individual officers is that, even if this approach did succeed in sustaining many complaints, it turns these lowest-ranking members of the department into scapegoats for larger organizational problems. Law professor Barbara Armacost argues that "it is unfair to lay the moral responsibility for police misconduct solely at the feet of individual officers."[457]

There is also no evidence that civilian review boards are more effective than traditional police internal affairs units in deterring police misconduct. In fairness, it must be pointed out that there is precious little research on the effectiveness of citizen oversight agencies generally, including both civilian review boards and police auditors. For that matter, there is precious little research on police internal affairs units.[458] Specifically, there is no evidence that one approach to internal investigations is more effective than another, or that one department is more effective than other departments.

Police Auditors in Action: Case Studies

The Special Counsel to the Los Angeles Sheriff's Department

The Special Counsel to the Los Angeles Sheriff's Department is arguably the most successful citizen oversight agency in the United States. During its first 11 years, it has established a very creditable record of investigating a wide range of issues related to police accountability, identifying problems that need correcting, making recommendations for change, and then monitoring the implementation of those recommendations. Perhaps most important, the Special Counsel's office is the one citizen oversight agency that has been able to present tangible evidence of its effectiveness.[459]

The Special Counsel operates as an independent agency under a series of fixed term contracts with the Los Angeles County Board of Supervisors. He hires his own staff, conducts investigations, and publishes semiannual reports. By mid-2004, there were 17 reports, which collectively provide a more detailed picture of LASD operations than of any other law enforcement agency in the United States.[460] Merrick Bobb, who began his career in police oversight on the staff of the Christopher Commission, has served as the Special Counsel since its inception, and was also the founder and director of the Police Assessment Resource Center (PARC), a private nonprofit organization that has emerged as an important organization in the field of police accountability.[461]

The original mandate of the Special Counsel was to reduce the costs of civil litigation over misconduct by the Sheriff's Department, which had been identified by the 1992 Kolts investigation as a serious problem.[462] Bobb has taken a very broad interpretation of the Special Counsel's mandate and addressed a wide range of issues within the LASD. As Box 6.1 indicates, he has investigated more than 25 different issues in six general areas (the exact number and the categorization are somewhat arbitrary because many issues overlap). Most issues have been reexamined several times, and some particularly important issues have been addressed in almost every report. Civil litigation, for example, has been addressed in 12 of the 17 reports. Some aspect of use of force (including both deadly force and physical force) has been addressed in 11 of 17 reports. As we shall see in our discussion of failed auditor offices, some auditors failed to use their authorized powers and consequently failed as accountability mechanisms.

Control of the Canine Unit. The best example of the effectiveness of the Special Counsel is its continuing attention to the LASD canine unit, which has been examined in nine separate reports. The 1992 Kolts Report expressed concern about the "numerous lawsuits" against the LASD involving the canine unit, alleging both excessive force and racial discrimination.[463] The Special Counsel's first report in 1993 found that the LASD had already taken steps to address the five problem areas identified by the Kolts Report. Dogs in the canine unit were now trained to "find and bark" rather than "find and bite." Handlers were no longer permitted to release their dogs without any announcement or warning to suspects. Announcements were also now being made in Spanish and English (an obvious issue in the heavily Latino Los Angeles area). Commanders now routinely reviewed all canine unit deployments with special attention to the "bite ratio" (the number of bites per total canine unit deployments) and investigated further whenever there were bites in more than 30% of all deployments. Released dogs were now immediately called off when it

Box 6.1   LASD Special Counsel Issues Investigated, 1993–2004

Issue	*Semiannual Report Number*
I. Use of Force Issues	
Canines	1st, 2nd, 4th, 5th, 6th, 9th, 11th, 12th, 15th
Force tracking (PPI)	1st, 2nd, 3rd, 4th, 6th, 7th, 15th, 16th, 17th
Data integrity	6th, 7th, 11th, 12th, 16th
Off duty incidents	7th
Pursuits	16th
Shootings	1st, 2nd, 3rd, 4th, 5th, 6th, 13th, 14th, 16th
II. Accountability Issues	
Civil litigation	1st, 2nd, 3rd, 4th, 5th, 6th, 7th, 9th, 11th, 13th, 14th, 15th
Risk management	3rd, 5th, 7th, 13th, 15th
Accountability (generally)	1st, 2nd, 3rd, 4th
Century Station	9th, 15th
Citizen complaints	2nd, 4th, 6th
Corruption	1st, 2nd, 13th
Office of Independent Review	14th
Racial profiling	13th
III. Training Issues	
Training (generally)	1st, 2nd, 6th
Force-related training	5th, 9th, 11th, 16th
FTO training	2nd, 3rd, 5th
IV. Personnel Issues	
Status of women	2nd, 3rd, 4th, 5th, 6th, 7th, 9th
Sexual orientation	6th
General personnel issues	1st
Promotions	7th, 9th

(Continued)

Box 6.1 (Continued)

Issue	Semiannual Report Number
Recruitment and hiring	3rd, 4th, 5th, 6th
Psychological services	2nd, 3rd
Sexual harassment	2nd, 11th, 13th
V. Los Angeles County Jails	
Jail conditions	5th, 6th, 7th, 8th, 9th, 17th
Medical care	12th, 13th
F. Other Issues	
Citizen Advisory Committee	1st
Committee	
Civil Service Commission	1st, 2nd, 3rd, 4th
Community policing	12th
Experiments	15th
Ombudsman's Office and Judges Panel	1st, 2nd, 3rd, 4th
Overview	10th

became apparent that the suspect was not armed. Finally, the LASD incorporated the use of canines into its use of force scale and placed it high on the scale just below use of deadly force.[464]

Subsequent Special Counsel reports documented attention to other aspects of the canine unit. The deployment of canines in auto theft cases was discouraged because many suspected auto thefts are actually juvenile joy rides that pose little danger to officers. In addition, new policies encouraged handlers to maintain close proximity to their dogs to keep more control over them. The Special Counsel's *11th Semiannual Report* found that several officers whose performance had been subject to stricter review had left the canine unit to "seek new challenges elsewhere" in the department.[465] This suggests that the new policies encouraged the departure of officers who realized they could not comply with the departments new standards of accountability.

To put the LASD canine unit in perspective, it should be noted that use of the canine unit was a major source of controversy in Cincinnati and that a consent decree with the U.S. Department of Justice required a series of administrative controls similar to those adopted by the LASD, including the "find and bark" standard.[466] In short, the Special Counsel achieved through internal policy review the reforms that a riot and a federal law suit were required to achieve in Cincinnati. The LASD's data on its canine unit provide impressive evidence of the positive impact of these reforms (Table 6.2). In 1991, the canine unit was deployed 1,228 times, with 213 "finds" and 58 bites. By 2001, there were 680 deployments, 185 finds, and only 22 bites. The number of bites, in fact, had reached an all-time low of only 7 in 1998.

Litigation Costs. The original mandate of the Special Counsel was to reduce the costs of civil litigation against the LASD. The *14th Semiannual Report* (October 2001) presented data indicating some success in this regard. Between fiscal years 1992–1993 and 2000–2001, the current docket of pending excessive force cases declined from 381 to 102 (and had reached a low of 70 in 1998–1999). Newly filed use of force cases declined from 88 in 1992–1993 to 67 in 2000–2001 (with a low of 41 in 1998–1999). Settlements in use of force related cases had also declined. The Kolts Report had found an average of about $5 million per year between 1989 and 1992; in the 2-year period including fiscal years 2000 and 2001 the settlements averaged $3.7 million a year.[467]

Official data on civil litigation are extremely problematic, however, and as typically reported cannot serve as a reliable performance measure for a law enforcement agency. There is always a considerable time lag between the original incident and the final settlement. The LASD paid out $27 million in use of force-related cases in fiscal year 1998–1999, for example, but about $20 million of that was for one 1989 case.[468] Thus, the money paid out in a given fiscal year represents incidents that are typically many years old and does not reflect current policies and practices of the department. In addition, one extreme case can grossly distort the trend data. A more sophisticated approach to reporting litigation costs would assign cases and costs to the years in which the original incidents occurred. In addition, it is advisable to distinguish between the costs associated with types of cases: for example, fatal shootings, excessive force, pursuits, routine property damage, and so on. To date, however, no one has developed a system for appropriately analyzing both the trends and the causes of litigation costs.

Table 6.1   Performance of the LAPD Canine Unit

Year	Deployments	Finds	Bites	Find–Bite Ratio	Ethnicity	
1991	1,228	213	58	27%	African American	23
					Latino	24
					Anglo	9
					Other	2
1992	1,030	225	51	22%	African American	13
					Latino	30
					Anglo	6
					Other	2
1993	940	179	42	23%	African American	22
					Latino	13
					Anglo	6
					Other	1
1994	921	183	45	24%	African American	19
					Latino	18
					Anglo	7
					Other	1
1995	840	151	31	20%	African American	14
					Latino	12
					Anglo	3
					Other	2
1996	708	121	15	12%	African American	5
					Latino	9
					Anglo	0
					Other	1
1997	734	115	10	8.7%	African American	3
					Latino	6
					Anglo	1
					Other	0

(Continued)

Table 6.1 (Continued)

Year	Deployments	Finds	Bites	Find–Bite Ratio	Ethnicity	
1998	626	84	7	8.3%	African American	1
					Latino	5
					Anglo	1
					Other	0
1999	539	88	15	17%	African American	7
					Latino	8
					Anglo	0
					Other	0
2000	569	152	19	12.5%	African American	8
					Latino	10
					Anglo	2
					Other	1
2001	680	185	22	11.9%	African American	8
					Latino	10
					Anglo	2
					Other	2

The Case of the Century Station. Chapter One of this book opened with a description of the Special Counsel's investigation of the troubled Century Station. There is no need to repeat the details here, but it is important to highlight the overall role of the Special Counsel on this issue. To summarize, the Special Counsel's initial investigation found that the high number of officer-involved shootings was not the result of a few bad officers or even a natural product of the social conditions of the neighborhood.[469] Instead, Merrick Bobb and his staff concluded that they were the product of serious management deficiencies: the concentration of young and inexperienced officers who were supervised by young and inexperienced sergeants;[470] and a ratio of sergeants to officers that exceeded the LASD's own recommended standard of 1 to 8, at times reaching the level of 1 to 20 or even 25.[471]

Exemplifying the new accountability, the Special Counsel's report focused on organizational change. As a result, a number of notable administrative changes were made in the Century Station. New commanders were assigned to the station; lieutenants spent more time in the field, personally responding to serious force incidents; greater use was made of positive reinforcement rather than punitive discipline. The result was a dramatic decline in shootings.[472]

Possibly even more important than the initial report was the follow-up report in 2002. When Bobb and his staff revisited the Station, they found that many of the old problems had reappeared. After officer-involved shootings dropped substantially in 1999 and 2000, they returned to their old high levels in 2001.[473] Shootings and other problems reappeared because the LASD failed to maintain the most important administrative reforms. The Special Counsel noted that "The captains and other supervisors responsible for the improvements in the Century Station had moved on," and their replacements did not maintain the same level of attention to shootings.[474] The lesson of the Century Station experience is how fragile reforms can be in policing. Improvements were undermined not by evil intent but by a combination of neglect, a failure of top commanders to ensure continuity, and to a certain extent the silent operation of traditional personnel procedures. The erosion of reforms in the Century Station probably offers a general lesson for all police departments. Perhaps the most important lesson is that role the Special Counsel can play, through its institutionalized auditing function, to monitor important accountability related reforms and ensure that they are maintained.

Monitoring the LASD's Early Intervention System. The fragility of significant reforms and the consequent need for continuous oversight are also evident in the Special Counsel's reports on the LASD's early intervention system, the Personnel Performance Index (PPI). As Chapter Five explained, early intervention systems are the pivotal mechanism in the new police accountability. For years the LASD's PPI was widely recognized as perhaps the best EI system in the country, and cited as a model for other department to emulate.

The Special Counsel's 2003 report on the PPI, however, found that it was not operating effectively. As already explained in Chapter Five, the report found that officer performance data were not being entered in a timely fashion, many reports were incomplete or with errors, and some commanders were not aware of the capabilities of the PPI or even its existence. The Special Counsel's report was extremely embarrassing to the LASD and, presumably, provoked actions to bring the PPI back up to its full potential.[475]

Foot Pursuits. The Special Counsel's *Second Semiannual Report* in 1994 identified a disturbing pattern of officer-involved shootings related to foot incidents.[476] Bobb followed up on this issue in 2003 with an explosive report finding that almost a quarter (22%) of all shooting incidents between 1997 and 2002 involved foot pursuits. Unlike vehicle pursuits, which must be documented in official reports, the department had limited data on foot pursuits. Bobb's report cited several cases of "tactically or strategically unsound" behavior by officers that put them in needless danger. In one, a deputy pursued a suspect alone, in the dark, in a high-crime area. He had lost sight of the suspect, and no one else in the department knew the officer's whereabouts. These actions involved violations of several department policies regarding foot pursuits. It also found that department supervisors were very reluctant to criticize officers who engaged in "tactically reckless" foot pursuits. Moreover, the suspect's original violation was running a stop sign—in short, not presumptively a dangerous felon. Bobb found this case typical of a common pattern of tactically reckless pursuit behavior, including single-deputy pursuits, partner splitting, failure to call for assistance, pursuing armed suspects who have disappeared from sight, poor communication among deputies, and pursuits with no plan of action.[477]

Bobb concluded his analysis with the recommendation that the department adopt the foot pursuit policy of the Collingswood, New Jersey, Police Department. This policy is discussed in detail in Chapter Four.

The foot pursuit report provoked an intensely hostile reaction from officers in the department—far more hostile than to any issue discussed in previous Special Counsel reports. The criticisms of officer behavior in foot pursuits touched a raw nerve related to the values of the police culture, wherein flight by a citizen or suspect is regarded as a direct challenge to an officer's authority (a variation on the "contempt of cop" theme that is associated with much officer use of force). Even though Bobb's report documented the extent to which many of these pursuits put officers in danger and were hardly justified by the minor nature of the suspected offense, the police culture clearly values zero tolerance of this form of perceived disrespect. In the end, Bobb's report revealed the extent to which much problematic police behavior is rooted in the police culture, and consequently the importance of changing the norms of that culture.

The LASD Office of Independent Review (OIR). In 2001 Sheriff Lee Baca of the LASD took the extraordinary step of creating the Office of Independent

Review (OIR), giving the LASD a second auditor paralleling the work of the Special Counsel. Staffed by seven attorneys, it has much the same mandate as the Special Counsel to examine the LASD's policies and procedures, make recommendations for change, and make public reports about its findings. The initial director of the OIR was Michael Gennaco, a former Assistant U.S. Attorney with experience prosecuting police misconduct cases. The OIR operates under contract to the LASD, and staff members are technically not department employees.

With an open-ended mandate, the OIR developed a methodology that often begins with a specific incident: a citizen complaint, a lawsuit, or some other indication of a problem. OIR staff review the quality of the LASD's investigations for thoroughness and fairness and look for policy failures that need correcting. By the end of its second year, the OIR had reviewed more than 300 cases.[478] As discussed earlier in this chapter, the OIR publishes summaries of its reviews of cases, including information on the disciplinary actions ultimately taken, providing an unprecedented degree of openness and transparency for the department.[479]

The review of individual cases often points to larger policy issues that need to be addressed. In one of the first important investigations, the OIR began with a false arrest suit involving an officer assigned to the LASD's community policing program. It found that officers in the program essentially operated with minimal day-to-day supervision and no clear policy about surveilling citizens. Many officers named in the lawsuit had never been trained in proper surveillance techniques. Also, the officers would gather in a fast food store parking lot and decide what to do that night. As a result of the OIR's report, the LASD developed a new policy on surveillance and provided training for COPS (community-oriented policing) unit officers, conducted in part by OIR staff.[480]

The OIR experience also illustrates the importance of police auditors as an institutionalized oversight mechanism. As it explained, " [the] OIR has been present long enough to have seen internal investigations go through the entire cycle, from inception through the decision-making and grievance process." As a result, the staff "has begun to learn the challenges and influential circumstances that can affect outcomes differently at the different stages. In particular, OIR has created checks to ensure that the principled decisions are not undermined by last-minute developments or hasty settlements that are reached behind closed doors."[481] In addition to identifying problems and undertaking investigations, the OIR "rolls out" to individual force incidents. An OIR attorney is on-call at all times and receives immediate notification of major force incidents. Upon arriving at the scene, the OIR attorney seeks to determine as quickly as possible whether the incident involves potential problems that merit further attention.[482]

In another important finding, the OIR also discovered that in a number of misconduct cases the LASD did not follow through and actually implement the disciplinary action it had decided on. The department had a practice of reaching "settlement agreements" in cases that called for some discipline short of termination. These agreements with the employee often include a requirement of remedial action such as alcohol counseling, anger management, leadership school, retraining, and so on. The OIR discovered, however, that the LASD had no effective system for ensuring officer compliance with the agreed-on terms. An audit of 19 randomly selected settlement agreements found that nine had either no compliance with the required remedial action or no documentation of compliance.[483] The OIR's finding on this point is paralleled by the finding of the Philadelphia Police Department's Office of Integrity and Accountability (OIA) that many officers were never disciplined for complaints the department had sustained.[484]

Summary: The Los Angeles Sheriff's Department. To sum up the activity of the Special Counsel and the Office of Independent Review, no other law enforcement agency in the United States is subject to such close scrutiny by outside investigators as is the LASD. What is particularly significant is that both auditors have found serious problems in a variety of different aspects of the department's operations and, more seriously, the recurring problem of the department's failure to follow through on actions and programs it has undertaken. This latter point highlights the need for continuing external oversight.

The San Jose Independent Police Auditor

The San Jose Independent Police Auditor (IPA) is an independent municipal agency with three major responsibilities: to receive citizen complaints, to monitor complaint investigations, and to promote public awareness of the complaint process.

With respect to the citizen complaint process, the IPA has the power to disagree with the findings of police internal investigations. Disagreements are rare, however. The IPA disagreed with the findings in 5% of the cases (a total of 6 cases) in 1999 and 2% of the cases in 2000.[485] The reasons for disagreement are instructive. In Case #1, the IPA disagreed because the internal investigation did not give sufficient weight to the testimony of an independent witness that corroborated the complainant's statement. In Case #3, the IPA held that the internal investigation failed to give sufficient weight to a witness officer's testimony that the subject officer had exacerbated the situation and provoked the hostile behavior of the complainant.[486] These and other cases are classic examples of the traditional police practice of ignoring or devaluing testimony supporting complainants' allegations. To the extent that a police auditor corrects this practice, the office serves as an important quality control over internal complaint investigations. It is also safe to assume that the IPA's monitoring deters many potential lapses by investigators. In short, the value of the IPA's monitoring activities is not measured by the percentage of disagreements per year, but by the process of ongoing monitoring and quality control.

Policy Review. The policy review activities of the IPA represent its most important achievements. By 2004 it had made a total of 95 policy recommendations; only seven were not accepted by the San Jose Police Department, although several were still pending.[487] The significance of these policy recommendations is not immediately apparent to all observers. Most of them involve relatively minor and insignificant "housekeeping" items. But therein lies their importance. The sustained attention to small issues has two important effects. First, addressing small issues prevents them from developing into larger problems. Second, the process of sustained oversight by an external agency socializes the police department into the habit of answering to outsiders on a regular basis. More than any single policy change, this process has the potential for changing the culture of a police department with respect to accountability.

The IPA's policy recommendations fall into several general categories. One group involves the *procedures for handling citizen complaints.* The IPA, for example, has recommended the following:

- Changes in the way complaints are classified
- Standardized definitions of complaint categories
- A new and more private room for interviewing complainants and witnesses
- New procedures to eliminate bias in investigations

- An explicit statement that retaliation against a citizen for filing a complaint or expressing a desire to file a complaint is forbidden
- A receptionist in the police department lobby to provide information and assistance to citizens
- More convenient access to restrooms for people waiting in the police department lobby
- Timetables for completing investigations

As Chapter Four explained, there have been problems related to all of these issues in departments across the country over the past 40 years. Each of these recommendations is designed to create a complaint review process that is more open and accessible to citizens and that processes complaints in a thorough and efficient way.

A second category of recommendations involves *police interactions with the public.* The IPA has recommended the following:

- An "on-lookers" policy explicitly stating that citizens have a right to observe police actions involving other citizens
- Closer supervision of strip searches of persons arrested for misdemeanor crimes
- A policy requiring officers to provide a prompt explanation or apology to citizens in cases in which a house had been searched by mistake
- A policy requiring officers to explain the reason for a traffic stop, search, or detention
- Development of a uniform definition of a racially motivated traffic stop or other contact

A third category of recommendations involves *police personnel issues.* The IPA has recommended the following:

- Improved training in communication and interpersonal skills
- Revised job descriptions that include skills specifically related to community policing
- Continued effort to employ bilingual officers
- Recruiting strategies to address family related issues

In the context of controversies over fatal shootings and egregious uses of force, many of these issues may appear somewhat trivial. Access to public restrooms in the police headquarters lobby hardly ranks in importance with fatal shootings of citizens. For the person who has come to police headquarters, however, the lack of a restroom is a very important issue. Failure to provide a restroom or a receptionist who can readily answer questions is likely to be a source of great frustration for citizens. Providing such services communicates to a key audience—people who have come to police headquarters either on their own or by compulsion—that the department cares about them as people. In the normal rush of business in a police department, which typically involves one crisis after another, it is easy for the chief and other command officers not to see such little issues and appreciate their significance.

At the same time, the policy recommendations related to police interactions with citizens involve some of the most sensitive and potentially explosive issues in policing. A strip search is a highly demeaning experience for most people, and it is imperative that the police conduct them only in the most urgent situations and with great care. Traffic stops are the most common form of contact between citizens and police officers, and simply explaining why a driver is being stopped can go a long way toward alleviating the inevitable resentment that many people feel.

The sheer breadth of the IPA recommendations is impressive, and it parallels the wide range of issues covered by the LASD Special Counsel.

Update on Previous Issues. As is the case with the Special Counsel to the LASD, the San Jose IPA has the capacity to revisit issues and publicly report on the status of prior recommendations. The *2001 Year End Report,* for example, devotes ten pages to "Updates on Prior Issues."[488] On the issue of police officials providing prompt and accurate information about where citizens could file complaints, calls to three different SJPD phone numbers yielded inaccurate or only partly accurate responses. Consequently, the IPA recommended a training program for all SJPD officials about the complaint process. With respect to shootings of citizens, the IPA reported that the SJPD had implemented the prior recommendation to enhance officer training and to provide officers with a wider range of less than lethal force options. The IPA noted that the number of shooting incidents had declined, perhaps as a result of these changes.

Community Outreach. The IPA also engages in an extensive program of community outreach. The *2001 Year End Report* devotes nine pages to community outreach, describing presentations to community groups, particularly youth groups, new police officers, and national associations. In 2002 the IPA published a very readable booklet for young people discussing both their rights and responsibilities when dealing with police officers.[489]

Summary: Impact of the IPA. Unlike the case of the canine unit in the LASD, however, no statistical evidence documents the success of the various San Jose IPA activities. Although the list of policy recommendations is impressive, we do not at this point have data or independent research indicating their impact on police operations. This is a serious shortcoming with respect to the entire new police accountability. Perhaps the most important issue for the future is the development of research methodologies to assess the impact of the various strategies and tactics discussed in this book.

Other Auditor Offices

The Boise Ombudsman

The Boise, Idaho, Ombudsman is particularly notable for its detailed follow-up reports on controversial incidents. In one notable case, the police officers responding to sexual assault allegations by two teenage girls engaged in extremely unprofessional and inappropriate behavior. They inappropriately concluded that the girls were lying, threatened to arrest them for perjury, leaked information about the incident to neighbors, and conducted interrogations without a third party witness in violation of departmental policy. The Ombudsman's report reviewed the facts of the case in detail, identified the specific areas of misconduct, and highlighted areas in which the police department needed to take corrective action.[490] In addition to the specific recommendations for reform, this and other Ombudsman reports serve to clear the air by providing an independent public examination of controversial police cases.

The Ombudsman staff also conducts immediate "roll outs" to shootings and other critical incidents. As noted in Chapter Three, this practice is now found in several other jurisdictions. The presence of an independent investigator on the scene of an incident is designed to preserve all relevant evidence and prevent blatant cover-ups of officer misconduct.

The Seattle Office of Professional Accountability (OPA)

The Seattle Office of Professional Accountability (OPA) occupies a unique and in many respects difficult role. The OPA director is not a sworn officer and commands the police department's internal affairs unit—in short, an insider–outsider. The most notable accomplishment of the OPA has been in leading a continuing program of review of use of force and racial profiling issues. A 2003 report on allegations of racial profiling by the Seattle Police Department was both detailed and extremely illuminating. After a careful review of the citizen complaint files, it concluded that a pattern of racial bias did not exist.[491] Particularly interesting was the report on the complaint against "Officer A" alleging racial bias:

> Officer A works in the traffic unit. He is a white male. In less than one month, he received two complaints from citizens he had ticketed who alleged bias. Complaint # 1 was from a black male who received three traffic citations from Officer A in a 15-minute period. The first citation was for going 51 mph in a 30 mph zone. After the stop, the now angry complainant revved his car engine and made an illegal U-turn maneuver in the presence of the issuing officer. He was stopped a second time and issued a citation for the illegal U-turn. The complainant then told the officer that he was going to continue to do U-turns and again revved his engine, spun his rear wheels, and made an illegal U-turn in front of oncoming traffic. This time, the complainant was cited for negligent driving.[492]

As with the Boise Ombudsman reports, this review and its publication on the department's Web site helps to clear the air about an issue that had caused great controversy in the community. But the OPA's inquiry into the racial profiling issue did not stop there. It has involved further efforts to investigate policies and procedures related to possible racial bias, all of which are described in public reports posted on the department's Web site.

Philadelphia Integrity and Accountability Office

The Integrity and Accountability Office (IAO) in the Philadelphia Police Department was created as part of a settlement of a suit by the NAACP, the ACLU, and the Barrio Project. Staffed by Ellen Ceisler, the IAO has made a practice of identifying particular issues for intensive investigation and public reports. Initial reports included *Use of Force* (1999), *Disciplinary System* (2001), and *Enforcement of Narcotics Law* (2002).[493] The reports on discipline and use of force are cited extensively in Chapter Three and need not be repeated here. Suffice it to say that they are especially valuable in the way they dig deep into the operations and culture of the Philadelphia department to provide a picture of how use of force incidents and disciplinary matters are handled. They reveal the extent to which formal policies, such as use of force reporting, can be undermined by subtle day-to-day practices, including the failure to complete reports, the failure to review reports and draw the obvious conclusions, and the failure to impose discipline in an obviously appropriate manner. Even more important, these reports indicate how these daily operational failures reflect larger management failures, which in turn reflect an organizational culture that is disinterested in accountability. In short, the IAO's investigations not only explain the underlying causes of the persistent accountability problems in the Philadelphia Police Department, but provide a direction for meaningful organizational reform.

SUSTAINING REFORM

One of the greatest problems in achieving police reform has been to sustain particular reform efforts and to ensure that new policies and procedures continue to function. The recommendations of numerous blue-ribbon commissions over the years have often never even been implemented. And where recommended changes were implemented, there is a long history of a failure to sustain them over the long-term and build a department that is truly "self-policing." In a comment that applies to 100 years of police reform efforts, Merrick Bobb observed that

> There is a daunting history of failure that precedes [these reforms]. Indeed, for as long as we have been monitoring the LASD, our hopes about permanent improvement at Century have risen and been dashed in regular cycles as each new Captain comes in full of energy and ideas and eventually leaves with the view that the job required him to push the same stone up hill year after year, only to see it roll back down again.[494]

In this regard, the saga of corruption-control efforts in the New York City Police Department is particularly noteworthy. The NYPD was struck by a major corruption scandal in 1969 (often referred to by the names Serpico or Knapp). In the wake of the scandal, Patrick V. Murphy was hired as Police Commissioner and proceeded to implement a number of administrative controls designed to limit corrupt practices. Murphy's reforms have been extensively covered in both popular and scholarly literature and are the basis for his reputation as one of the preeminent police chiefs of recent decades. In brief, the centerpiece of his program was to create a decentralized procedure for corruption control. Field Internal Affairs Units (FIAUs) handled the bulk of allegations of misconduct, whereas a centralized Internal Affairs Division (IAD) would take responsibility for major cases.[495]

The 1994 Mollen Commission report, which was prompted by a subsequent corruption scandal, found that the accountability mechanisms established by Murphy had essentially "collapsed." Several factors contributed to this collapse. The entire system relied too heavily on the personal commitment of Commissioner Murphy and was never adequately institutionalized. The FIAUs did not receive sufficient resources to accomplish their goals. And most important, commanders willfully turned a blind eye toward inappropriate police behavior. In short, there was no institutional commitment to integrity within the department and no external agency to identify and publicize the growing problems.[496]

The fate of reforms in the Los Angeles Police Department following the 1991 Christopher Commission report is another story of failure. A 5-year follow-up report found that many recommendations had not been implemented. Two years later, the chief of the LAPD announced with great fanfare that virtually all of the Christopher Commission recommendations had been implemented. But in early 1999, the LAPD was engulfed in the enormous Rampart Scandal, which generated no fewer than three reports on what went wrong. Most alarming was the LAPD's own Board of Inquiry report, which concluded, among other things, that the personnel evaluation process in the department was widely regarded as worthless. In short, it appears that the accountability mechanisms in the LAPD failed during a time of great public pressure for reform and while the department was congratulating itself on improving.[497]

The failures of reform in New York and Los Angeles should not be interpreted as special cases. Instead, they highlight a pervasive problem in policing. The simple fact of the matter is that changing a police organization is an extremely difficult task. In addition to the inherent inertia of all bureaucracies, policing is heavily shaped by informal officer norms at the street level. Eliminating deeply ingrained practices and habits of mind throughout a police bureaucracy is an enormous challenge. Although virtually every expert on policing will say that the key is "leadership," few of them can offer specific guidance on how a strong leader can effect lasting change.

The auditor model represents one way to institutionalize reform. The special virtue of this approach is that it involves a permanent external oversight agency with the capacity not only to recommend changes in police department policies and procedures, but to conduct follow-up investigations on the implementation of prior recommendations.

If anything, the reports of the Special Counsel to the Los Angeles Sheriff's Department provide sobering confirmation of the problem of effecting change in a law enforcement agency, and the consequent need for continuous external monitoring. As Box 6.1 indicates, the Special Counsel has reexamined most issues several times over the course of 10 years. Although the Semiannual Reports document genuine improvement in many areas (e.g., the canine unit), the reports are also filled with examples of recommendations that have not been implemented, effective reforms that were allowed to lapse (the Century Station, the PPI), or new aspects of old problems (foot pursuits).

THE LIMITS OF THE AUDITOR MODEL

Although the auditor model of citizen oversight of the police has enormous potential for effecting change in police organizations, it also has a number of weaknesses and is hardly a panacea for police problems. These limitations of the police auditor model fall into two basic categories. There are some weaknesses inherent in the basic concept. There are also some notable examples of police auditors that have simply failed.

Inherent Limitations

Perhaps the major limitation of the police auditor concept involves public credibility: the perception by some community leaders that the auditor is not truly independent of the police department. This author had several debates with John Crew, former Director of the ACLU Police Practices Project in San Francisco, over his criticisms of the San Jose auditor and the auditor model in general.[498] Portland Copwatch, the most vocal watchdog group in Portland, Oregon, has been a fierce critic of the auditor model in that city.[499] The primary criticism of the auditor model is that it does not investigate individual complaints. Among many activists there is very deep distrust of any system that leaves the police in charge of complaint investigations. As a consequence, these activists view the auditor model as a weak form of oversight.

The criticisms of community activists are important because *public perception* is such a crucial aspect of police accountability. It is not enough that a police department reduce the number of excessive force incidents and investigate complaints thoroughly; it is also necessary that the department be perceived as doing so. As this book has already suggested in several places, the focus on investigating individual complaints on the part of community activists is misplaced. The independent review of individual complaints on a case-by-case basis is unlikely to address the systemic organizational problems that are at the heart of police misconduct. The very essence of the new accountability is organizational change, particularly through the identification of patterns of misconduct and the development of corrective policies and procedures.

The work of a police auditor is particularly unappealing to many community activists whose organizing style includes a preference for confrontation and drama. The work of a successful police auditor is necessarily undramatic: long and detailed analyses of the nitty-gritty details of the inner workings of a police department's response to on-the-street behavior and alleged misconduct, such as one finds in the reports of the Philadelphia Integrity and Accountability Office or the LASD Office of Internal Review. These reports are rich material for policy makers and scholars (certainly for the author of this book), but not for many community activists.

Another criticism of police auditors is that policy recommendations are merely recommendations and can be rejected by the police department. This is certainly true. But it is also true that civilian review boards also only have the power to recommend disposition of complaints and not to compel discipline. (Giving review boards that power would require a significant change of local and state law and, in any event, is not recommended by many advocates of oversight, including this author.) As this chapter has argued, the policy review process represents a good way of developing an informed public debate over police policies. Although it is true that police auditors largely have only the power to expose problems and make recommendations, this power should not be underestimated. Given the historically closed character of American police departments, the power to dig deep into police departments, issue public reports, and conduct follow-up investigations has enormous value that should not be underestimated.

Limitations on the independence of police auditors from the police department and other political pressures are also potential problems. It is less of a problem in San Jose, where the IPA is a separately chartered city agency with a high degree of independence not only from the police department but from other elected officials as well. The Portland Independent Review Office, administratively housed in the City Auditor's office, is also highly independent of the police department. At the other extreme, both the Special Counsel and OIR in the Los Angeles Sheriff's Department function as contract employees and can be terminated relatively easily. Auditors in some other cities are housed under the mayor or city manager and are inevitably subject to political pressures from that office. The Seattle OPA is in a particularly ambiguous position, being part of the command staff but at the same time an outsider, expected to work closely with the chief (on the one hand this is an advantage), but at the same time expected to be an outside critic. It is not clear whether this balance can be maintained.

Failed Auditors

Some police auditors have simply failed to function effectively. The Seattle Police Auditor (a different office than the newer OPA), created in 1993 as a compromise over a demand for a citizen review board, is a particularly notable example. The auditor was charged with the responsibility of reviewing the files of the Seattle Police Department's (SPD) internal affairs unit and making periodic public reports. A retired judge was hired by contract to serve as the auditor. The Seattle auditor's reports are extremely short, containing very little detail about the complaint process or the SPD in general. In terms of content, they pale by comparison with the reports of the Los Angeles County Special Counsel, the San Jose IPA, and the Boise Community Ombudsman. In particular, they contain none of the policy recommendations found in these other reports and provide very little information about police activities.[500] The failure of the Seattle Police Auditor led to the creation of the OPA as an alternative strategy for achieving accountability.

The Seattle Police Auditor has been a very low-visibility office and has undertaken no community outreach activities. Nor did the Seattle auditor involve himself in oversight at the national level. Most other auditors and citizen oversight officials around the country reached out to their colleagues in other cities and often attended the annual meetings of the National Association for Citizen Oversight of Law Enforcement (NACOLE) and the International Association for Citizen Oversight of Law Enforcement (IACOLE). Through these activities they learned what other offices were doing and took new ideas home with them. The Seattle Police Auditor was so ineffectual that it was not even mentioned in the flurry of investigations and reports following a major scandal in 1998 that led to the creation of the Office of Professional Accountability. It is a commentary on the irrelevance of the Seattle auditor that it was not even mentioned in the final report of the mayoral-appointed Citizens Review Panel nor in the Seattle Police Department's lengthy and comprehensive accountability plan.[501]

The failure of the Seattle auditor was a result of two factors. First, the enabling ordinance did not direct the auditor to review policies and procedures as did ordinances in other cities. Second, the person hired as auditor simply chose to play as limited a role as possible *and received no direction to the contrary* from the responsible political officials. Thus, there was a failure of vision on part of all officials involved.

The Albuquerque Independent Counsel (IC), created in 1987 and abolished in 1998, failed to use powers that were expressly granted to it by ordinance. Created by city ordinance, the office functioned as an independent contractor with the responsibility for reviewing all citizen complaint investigations conducted by the police department's internal affairs unit. The language of the ordinance gave the IC full authority to direct the complaints process. The IC had the authority to reject investigations deemed inadequate and to request further investigation. An evaluation of the oversight mechanisms in Albuquerque (coauthored by this author) found that although the IC provided "an important measure of citizen oversight" of the police department, the individuals holding that office had "not fully utilized existing authority to review the policies and procedures of the APD and make recommendations for change." In addition, the Independent Counsel had played "no public role" in the community (and in fact, was not known by many community activists) and, as a consequence, had contributed absolutely nothing to building public confidence in the oversight process.[502] The individual who had been serving as IC deliberately kept a low profile and was not known by community leaders. As a result, the IC had not developed community trust in the office as an oversight agency. Also, the evaluation found that the IC had made very little use of its authority to conduct studies of police policies and make recommendations for change.

The Albuquerque evaluation concluded that the failure of the IC was due to a lack of "political will."[503] The formal powers of the IC were substantial. But it was evident that informal decisions had been made by political officials to have the IC do as little as possible to challenge the police department or create expectations in the minds of community activists. The important conclusion of the Albuquerque evaluation is that formal powers are necessary but not sufficient to create an effective citizen oversight mechanism. A serious commitment on the part of public officials—which the evaluation defined as political will—is also necessary.

THE CONDITIONS OF SUCCESS

In April 2003, all 12 police auditors met for a conference in Omaha, Nebraska. After intensive discussions of their respective roles and activities, they drafted a set of Core Principles for an Effective Police Auditor's Office. This document appears in Figure 6.2.

CONCLUSION

The auditor approach to citizen oversight of the police is rich with potential for achieving genuine improvements in policing. The assumptions and operations of police auditors are consistent with the other elements of the new police accountability described throughout this book. Particularly important is the emphasis on organizational change and the capacity of police auditors to revisit earlier issues and examine the implementation of prior recommendations.

It is, however, necessary to temper any judgment on police auditors with the qualifier, "potential." Although the Los Angeles Sheriff's Department Special Counsel and the San Jose Independent Police Auditor have very creditable records, we still do not have scientifically rigorous independent evaluations that provide convincing evidence of reductions in the use of excessive force and other forms of police misconduct over the long-term. Of course, we do not have that kind of evidence for any police reform measure, whether of the traditional variety or the new police accountability.

The police auditor concept does not automatically translate into success. Several factors are necessary for an effective auditor. Most important is a vision of police accountability and an understanding of what is necessary for the reduction of misconduct and a grasp of the role that an auditor can play in that regard. Any auditor's office that chooses a narrow interpretation of its role will not fulfill its potential. Auditors, of course, are a part of the political environment in which they work. What a police auditor does or does not do is determined in large part by the direction he or she receives from the larger

INDEPENDENCE

A police auditor's office must be fully independent of the law enforcement agency under its jurisdiction.
Specific language in the enabling ordinance must indicate that an auditor may be removed from office only for cause and through a clearly defined removal process.

CLEARLY DEFINED SCOPE OF RESPONSIBILITIES

The scope of the responsibilities of a police auditor's office must be clearly defined by ordinance (or contract).
Specific language, for example, must define the auditor's responsibility to audit complaint files, have unfettered access to all relevant records and reports, make policy recommendations, issue public reports, investigate individual critical incidents, and so on.

ADEQUATE RESOURCES

A police auditor's office must have adequate resources to ensure that all duties can be conducted effectively and efficiently.
Adequate resources primarily include full-time professional and clerical staff.
Part-time staff only are not considered adequate.
Volunteer staff are not adequate.
The exact size of an auditor's office staff should be based on a formula reflecting the size of the law enforcement agency under the auditor's jurisdiction, as measured by the number of full-time sworn officers.

UNFETTERED ACCESS

A police auditor must have unfettered access to all documents and data in the law enforcement agency.
This unfettered access must be spelled out in the enabling ordinance.
The only exception to this rule would be files related to an ongoing criminal investigation.
All documents must be provided to the police auditor without charge to the auditor's office.

FULL COOPERATION

A police auditor must have the full cooperation of all employees of the law enforcement agency under its jurisdiction.
All employees, including sworn officers, shall cooperate as a condition of their employment.
With respect to potential self-incrimination, the standards defined in *Garrity v. New Jersey* shall prevail.

SANCTIONS FOR FAILURE TO COOPERATE

The enabling ordinance of an auditor's office must specify sanctions for failure to cooperate with the work of an auditor on the part of any law enforcement agency employee.

(Continued)

Figure 6.2   Core Principles for an Effective Police Auditor's Office

---

PUBLIC REPORTS

A police auditor must issue periodic public reports.
Such public reports shall be issued at least once a year and, ideally, more frequently.

NO PRIOR CENSORSHIP BY THE POLICE DEPARTMENT

Reports by the police auditor shall not be subject to prior censorship by the law enforcement agency.
A police auditor may reject any and all demands by the law enforcement agency to see draft copies of public reports.

COMMUNITY INVOLVEMENT

A police auditor must have the benefit of community involvement and input.
Community involvement and input can best be achieved through an advisory board consisting of members who represent the diverse composition of the local population.

CONFIDENTIALITY AND ANONYMITY

The work of a police auditor must respect the confidentiality of public employees as defined in the applicable state statute.
Violation of confidentiality shall be considered a serious breach of professional standards.
In the interests of enhancing public understanding, a police auditor may report on specific incidents with personal identifiers removed without violating standards of confidentiality.

ACCESS TO THE POLICE CHIEF OR SHERIFF

A police auditor must have direct access to the chief executive of the law enforcement agency under its jurisdiction.
Upon request, a police chief or sheriff must agree to meet with the police auditor.
It is understood that a chief executive may decline to meet in the case of an unreasonable number of such requests. Failure to meet with a police auditor for a period of one year shall be considered unsatisfactory performance on the part of a chief executive and shall be taken into consideration in performance review.

NO RETALIATION

The enabling ordinance of an auditor's office must specify that there shall be no retaliation against the auditor for work done as a part of the auditor's responsibilities, including statements made in public reports.

Figure 6.2 (Continued)

environment. Two of the failed auditors are noteworthy in this regard. The evaluation of the Albuquerque Independent Counsel attributed its failure to a lack of "political will." The Independent Counsel himself and the political leaders who appointed and directed him simply chose not to push for an active agency. Similarly, the original Seattle Police Auditor failed because no one demanded that the office do more.

Assuming a proper vision and direction, a police auditor's office needs sufficient resources to do its job adequately. Many civilian review boards have been starved into failure by a lack of resources. Mike Gennaco, head of the LASD Office of Independent Review, observed that his office's initial success was made possible by "the commitment of significant resources by the [Los Angeles County] Board of Supervisors." The OIR began with a staff of six full-time attorneys, all of whom had backgrounds in civil rights and criminal law.[504]

Closely related to sufficient resources is the cooperation of the chief of police or sheriff. Gennaco reports that he enjoys "unfettered access to LASD materials."[505] Teresa Guererro-Daley has also enjoyed the full cooperation of a succession of chiefs of the San Jose Police Department. By contrast, the New York City CCRB has faced hostility and noncooperation from the NYPD over the course of several decades. The success of the San Francisco OCC has also been hindered by a lack of support from a succession of mayors and often-covert noncooperation from the San Francisco Police Department.

In the end, the issues of support from the political environment and cooperation from police chief executives apply to all forms of attempted police reform. If we have learned anything over the past several decades, it is that meaningful reforms are easily undermined, if only through quiet neglect. Federal courts can order a department to adopt a new use of force policy, but the department itself must implement it. And if the elected officials with direct responsibility for a police department do not carefully monitor implementation it is very likely that nothing will change. The police auditor is a potentially valuable tool for police accountability. Whether that tool will be used and used properly is a question that will be answered by people outside the auditor's office.

SEVEN

THE FUTURE OF THE
NEW POLICE ACCOUNTABILITY

————•⋅●⋅•————

The new police accountability is a dramatic development in American policing. It represents a coherent package of reforms that has the potential to effectively address historic problems of officer misconduct. Particularly important is that, unlike previous reform efforts, the strategies and tactics of the new police accountability have the potential to change the organizational culture of a police department and initiate a self-sustaining accountability effort. This best of all possible outcomes is by no means certain, but the potential is there.

This final chapter has three purposes. First, it summarizes the larger themes of the new accountability, and in particular how it improves over past reform efforts. Second, it discusses the obstacles facing successful implementation of the new accountability. It would be wishful thinking to ignore these obstacles and assume that progress is inevitable. In fact, there are very good reasons for skepticism. Finally, it offers some thoughts about how these obstacles can be overcome, including a discussion of some police departments that are "getting it right" in terms of proactive self-scrutiny with regard to accountability.

IMPROVING ON PAST REFORMS

The historic significance of the new police accountability, as explained in Chapter Two, is the way it addresses the limitations of past police reform efforts. It is important to revisit those limitations and highlight how the new police accountability may succeed where past efforts failed.

Controlling Day-to-Day Police Work

One of the great failures of many past police reforms is that they did not reach deep into police operations and affect day-to-day police work. Many of these reforms were well intentioned, but they were often formalistic and did not affect routine policing. Supreme Court rulings on police procedures enunciated grand principles of constitutional law but did not address the many ways police officers could evade their intent or the simple failure of departments to implement the requirements of major decisions. Raising educational standards for recruits and improving training are laudable and even necessary reforms, but in and of themselves they do not necessarily improve on-the-street police work for the simple reason that they can be undermined by more immediate countervailing pressures. In a similar way, an organizational chart that meets recognized standards of good management does not necessarily mean that the department will do what is necessary to hold officers on the street accountable. The organizational chart does not reflect the organization's culture—the working norms that shape day-to-day operations.

The potential of the new police accountability lies in the capacity of key strategies and tactics to penetrate police operations and control the behavior of officers on the street. The most important is the comprehensive use of force reporting system described in Chapter Three. It is now understood that this system must cover a growing range of officer activities posing risks to the lives, liberties, and safety of citizens: not just use of weapons, but vehicle pursuits, foot pursuits, use of canines, and other risky actions. The future of the new police accountability includes expanding the list of critical incidents covered by detailed policies and reporting requirements.

Although it is true that a number of important Supreme Court rulings spoke to on-the-street officer behavior, they did not include mechanisms to guarantee their implementation. The new police accountability addresses this crucial problem in several ways. The practice of conducting immediate "rollouts" to shootings and other critical incidents helps to ensure that incidents will be thoroughly and fairly investigated. Follow-up investigations to determine if an incident raises policy or training issues that need to be addressed can turn use of force reports into meaningful learning tools. Early intervention systems are particularly important in terms of identifying patterns of use of force by officers and initiating appropriate corrective action. Ongoing investigations by a police auditor, an independent oversight agency that is willing to be critical of the current department administration, can help ensure that all of the above-named procedures are functioning as intended. No one of these programs can guarantee that a use of force reporting system will achieve its purpose, but in combination they have the potential to alter the organizational culture of a department and create a culture that takes the reporting system seriously.

Enhancing Frontline Supervision

Past police reforms rarely addressed the critical role of supervisors, and street-level sergeants in particular, in controlling the conduct of officers under their command. The failure of reformers and the law enforcement profession itself in this regard is paralleled by the failure of the research community to study sergeants. We know very little about what sergeants do and what activities are most effective. These failures are all the more remarkable in light of the near unanimous recognition by all experts that street-level sergeants are the key to good policing. Robin Sheppard Engel's research on the styles of sergeants identified two styles that did not involve supervision in any meaningful sense. In one of those styles, in fact, supervisors defined their role in terms of protecting their officers from discipline by the command staff. This style is antithetical to any serious accountability program.[506]

The new police accountability enhances the role of frontline supervisors in several ways. Early intervention (EI) systems have tremendous potential for transforming the role of frontline supervisors. They give supervisors a database for assessing officer performance and emphasize, if not force, proactive supervision in terms of identifying potential problems and intervening early. The requirement (in some systems) that sergeants review the EI system database on a daily basis, meanwhile, creates a new standard of intensive supervision. Finally, EI systems provide a database that can be used to evaluate sergeants and hold them accountable. It is possible, for example, to determine how often a sergeant accesses the EI system database, to spot sergeants who have an unusually large number of officers who are identified by the EI system, and finally to assess whether the performance of officers who are counseled by a sergeant improves in subsequent months.[507]

In addition, the new standards for investigating use of force incidents include an evaluation of the actions (or inactions) of the supervisors in response to the events in question. Traditionally, investigations have focused on the individual officer who fired the shot or used a questionable level of force. Despite the fact that all experts acknowledge that the street sergeant plays a critical role in policing, the role of this person in critical incidents has not been examined as part of routine investigations. Finally, police auditors can continuously monitor the activities of sergeants to see if their activities conform to the expectations of the new accountability.

Proactive Efforts to Reduce Officer Misconduct

The new police accountability includes several programs that proactively attempt to reduce officer misconduct. In the past, police departments took a reactive stance toward misconduct incidents: An officer was disciplined after some act of misconduct or a new policy was developed only after a major misconduct crisis.

Early intervention systems are the principal proactive program. By analyzing patterns of officer performance, they seek to identify problematic behavior "early" and to take corrective action before it leads to a more serious incident. In this respect, EI systems represent a complete reversal of the culture of defensiveness and denial that has traditionally characterized police departments. As discussed in Chapter Five, members of police departments have always known who the bad officers are, but no systematic efforts were made to do something about those officers and their problems. An effective early intervention system, however, requires both a use of force reporting system and an open and accessible citizen complaint process that feeds into it timely and accurate data. This highlights the point that no single element of the new police accountability can stand alone. The different elements depend on and reinforce each other.

The investigation of use of force reports in the new police accountability also includes important proactive elements. It is now a recommended practice that use of force incidents be reviewed for the purpose of determining whether the incident calls for policy changes or training needs. Historically, incidents have been reviewed only to determine officer misconduct: whether the officer violated department policy, and, in very serious cases, whether there are grounds for criminal prosecution. It is also a recommended policy that a department conduct a full performance review of an officer involved in a serious misconduct incident (the EI system provides much of the raw material).

Finally, police auditors have the capacity to dig deep into police operations, identify shortcomings that permit misconduct to occur, and recommend the necessary changes in policy and procedure. The many policy recommendations by the San Jose Independent Police Auditor described in Chapter Five are particularly impressive. Although many involve seemingly small issues, the failure to attend to such small "housekeeping" items is often the underlying cause of major use of force problems.

Systematic Data Collection and Analysis

In the past, police departments simply did not know what their officers were doing. Without a comprehensive use of force reporting system, departments had no idea how often officers used force, or in what circumstances, by which officers, and against whom. The recent spread of traffic-stop data collection exposed the fact that departments did not have a full picture of traffic enforcement patterns, including who was being stopped and the outcomes of those stops. As a consequence, it was virtually impossible to identify specific performance problems, either in terms of the officers or categories of citizens involved. Nor was it possible to determine if reforms in policies and procedures had any impact on officer activities. Barbara Armacost, in an illuminating discussion of police organizational culture, points out that a problem associated with wrongdoing in organizations is "fragmented knowledge." Responsibility for different issues is parceled out in bureaucracies to the point where those with responsibility for primary operations (e.g., patrol in a police department) are isolated from those responsible for accountability (e.g., internal affairs). Key actors do not know what is going on in other sectors of the organization, and in some respects do not want to know.[508] The systematic collection and analysis of data has the potential to correct these problems, and for that reason is at the heart of the new police accountability. Early intervention systems are the centerpiece of this effort, but use of force reporting systems and professional citizen complaint procedures are important in and of themselves.

Permanent External Monitoring

Two of the great shortcomings of past reform efforts have been the failure to implement recommended reforms and the failure to ensure that reforms, once implemented, are maintained. Police auditors represent a practical and effective solution to this problem. As explained in Chapter Six, a police auditor is a permanent government agency with both the authority and the resources to conduct continuing investigations of police issues. The Special Counsel to the Los Angeles Sheriff's Department has set the standard by taking a broad license and investigating a wide range of issues within the department. As Chapter Six argued, one of the most important aspects of a police auditor's role is its capacity to revisit issues covered in previous reports and assess the extent to which recommended changes have been implemented.

APPLICATIONS

The strategies and programs of the new police accountability have several potential applications: for police departments, for researchers, and for the community at large.

The New Accountability as a Prescription for Reform

The strategies and tactics associated with the new police accountability serve as a prescription for police reform. Every department should have a comprehensive critical incident reporting system as described in Chapter Three, an open and accountable citizen complaint process as described in Chapter Four, and an early intervention system as described in Chapter Five. City councils, county boards, and state legislatures, meanwhile, should consider and establish some form of external auditing for their law enforcement agencies.

The model of reform described here is based on the current pattern of practice litigation. The various consent decrees and memoranda of agreement provide a list of best practices that departments should adopt. This book has simply amplified and expanded on the relatively short list of reforms found in each of the settlement agreements.

The New Accountability as a Self-Assessment Tool

In a similar way, the strategies and tactics of the new accountability can serve as a self-assessment tool for police departments. Rather than waiting to be sued for repeated use of excessive force, a department can proactively use the aspects of force reporting, citizen complaint procedures, and early intervention systems described in this book as a checklist to determine whether or not its practices are consistent with nationally recognized best practices. In practice, this means a department should ask itself, "Do we conduct immediate roll-outs to shooting incidents? Do we review all critical incidents to determine whether there are policy or training issues that need to be addressed?"

Indeed, whether or not a police department engages in such an ongoing, proactive self-assessment should be a major criterion for evaluating the quality of its management. Some examples of two police departments that are now doing this are described later in this chapter.

The New Accountability as a Research Tool

The strategies and tactics of the new police accountability also serve as a research tool and analytic framework for social scientists studying the police. They represent the issues that researchers should investigate and attempt to measure. In particular, the framework serves to move research efforts beyond official data on use of force and citizen complaints and toward investigating the processes by which those data are collected and processed. Several examples illustrate the research possibilities.

Comparing use of force rates in two departments, for example, is meaningless if those departments have very different procedures for responding to force incidents. One department may require reports on handcuffing, but the other does not; in one department, reports are not always reviewed by supervisors, whereas in the other all are reviewed; one department regularly audits its force reporting system, whereas the other does not. Researching these process issues will provide a more meaningful assessment of a department's commitment to accountability than the official data on use of force.

Along the same lines, an evaluation of a department's citizen complaint process could investigate whether potential complainants are turned away at precinct stations; the percentage of cases in which complaint investigators regularly fail to contact and interview witnesses; the percentage of cases in which the final disposition appears not justified by the facts presented in the investigative report; or the percentage of cases in which the discipline imposed appears to be lenient given the gravity of sustained allegation.

The importance of research cannot be understated. Candor compels us to point out that the discussion of the new police accountability in this book has been notably short of evidence based on independent research into whether the strategies and tactics discussed achieve their objectives. A recent comprehensive survey of police research by the National Academy of Sciences concluded that there is very limited evidence that written policies and reporting systems effectively reduce the misuse of force, for example.[509] The reduction of police misconduct and the improvement in police–community relations are urgent public policy issues. It is imperative that social science researchers investigate the various aspects of the new police accountability to determine what works and what does not with respect to achieving both lawful police activity and activity that is perceived to be fair and impartial by the public.

The New Accountability as a Resource for Community Activists

The strategies and tactics of the new police accountability also serve as a resource for community activists by providing them a specific list of the reforms they should seek. Too often, in the wake of an egregious shooting or excessive force incident, community activists mobilize to demand greater police accountability but are not familiar with the current best practices, and therefore do not make the proper demands. Their demands are often couched in vague terms of "citizen review" (without a specific sense of what a review board would actually do) or better training for officers (without a sense of how training translates into on-the-street performance). The new police accountability as described in this book provides a checklist for community activists in the same way that it serves as a self-assessment tool for police departments.[510]

AN UNCERTAIN FUTURE

Despite the great potential of the new police accountability for improving policing, it would be wishful thinking to assume that it will necessarily be implemented. The basic problem, which has bedeviled police reformers for generations, is how to effect lasting change in a large and complex bureaucracy such as a police department. Police history is filled with examples of promising reforms that did not last. It is sad—but necessary—to report that there are many examples of failings within the new police accountability itself. The following are a few notable examples of failed or inadequate reforms.

Examples of Failure

- The Integrity and Accountability Office in the Philadelphia Police Department still finds commanders who regard the basic use of force report and review requirement as a nuisance.[511]
- The Special Counsel to the Los Angeles Sheriff's Department found in early 2003 that the department's nationally recognized early intervention system—the PPI—was not functioning properly and that some commanders were not even aware of its capabilities.[512]
- More than 12 years after the initial Christopher Commission recommendation, the Los Angeles Police Department still had not implemented its early intervention system (initially Teams I and later Teams II).[513]
- In Oakland, Philadelphia, the Los Angeles Sheriff's Department, and Riverside, California, departments had failed to maintain their own official standards of having one sergeant supervise no more than seven or eight officers.[514]
- The LAPD's own internal report on the Rampart Scandal and the IAO in Philadelphia concluded that their respective departments' routine personnel evaluations were worthless.[515] Like the span of control principle, routine personnel evaluations were a basic element of the *old* police professionalism.
- In 2003 the Omaha Police Department offered no mandatory in-service training for its officers.[516]
- In the New Jersey State Police, 4 years after the consent decree was signed and in the context of regular review by the court-appointed monitor, the in-car videos of traffic stops were not fully usable.[517]
- In the Oakland Police Department, the Internal Affairs Division stopped receiving use of force reports at one point. Moreover, this occurred while the department was under the direct scrutiny of a court-appointed monitor, when one would expect all members of the department to be especially vigilant about complying with all tasks covered by the settlement agreement. The monitor subsequently found that the problem was due to a "clerical miscommunication" rather than any deliberate attempt to obstruct the use of force reporting system.[518]
- The Integrity and Accountability Office in Philadelphia identified some officers who were found guilty of misconduct were never disciplined. In the Los Angeles Sheriff's Department, the Office of Internal Review

found that some settlement agreements requiring officers to undergo some kind of remedial training or counseling were never enforced.[519]

Recurring Obstacles to Sustained Reform

The cases listed above represent several recurring themes in the failure of police reform. This section identifies the principal themes, discusses some of the underlying causes, and suggests a solution.

Failure to Adopt Basic Standards of Accountability

The hostility to procedures for requiring reports on critical incidents and having supervisors closely review them, which the IAO found in the Philadelphia Police Department, represents a failure to adopt a basic standard of accountability that has been accepted in principle in departments for more than 30 years. This reflects an organizational culture that is stuck in the past and has not kept pace with evolving professional standards. The failure of the Omaha Police Department to institute regular in-service training for all officers represents a similar failure to keep up with long-established national standards.

Failure to Maintain a Department's Own Standards

The failure of several departments to maintain their own standards regarding the proper ratio of sergeants to officers (the span of control) represents a failure of management to pay close attention to day-to-day operations. From one perspective, it is easy to understand how this happens. Police departments face continuous personnel problems. The combined effect of retirements, resignations, terminations, suspensions, vacations, and sick leave means that units are always facing shortages of available personnel. From the perspective of effective management, however, it is disturbing that some departments allowed themselves to fall far below their own standards for lengthy periods of time.

Example #1: The Problem of "Policy Drift"

Even after a good use of force and critical incident policy is in place, powerful forces are at work to erode compliance with that policy. Without close, ongoing supervision and regular in-service training, police practices on the street can easily begin to drift away from official policy. At some point, officers can begin to believe that their practice is official policy (e.g., as in "this is the way we do things"). The Omaha Public Safety Auditor calls this "policy drift."[520] Soon after taking office in 1991, Omaha Public Safety Auditor Tristan Bonn found that traffic stop and search practices by Omaha police officers often reflected the shared folklore among patrol officers and street sergeants about "how we do it." And on one point—the power to search passengers in traffic stops—informal practice had found its way into formal policy. The auditor found that the department's written policy stated that passenger searches were justified by state law. But no such law existed. The drift away from official policy (and in some cases state law) is abetted by the lack of ongoing training. The auditor found that Omaha Police Department—almost alone among big city departments—had no program of mandatory in-service training about traffic enforcement or any other subject.[521]

The process of policy drift is easy to understand if there is no regular system of policy review, in-service training, and external auditing. Departments are beset daily by pressures from all directions: public outcry about an increase in crime, demands from community groups, pressure from mayors and city council members for all sorts of small adjustments in practice, and so on. Even more serious are the daily crises that distract the attention of supervisors away from maintaining hard-won policy changes. As a result, actual practice on the street can easily drift away from the requirements of official policy.

Another source of policy drift is overt pressure from the rank and file, particularly regarding special units or tactics. Special Counsel Merrick Bobb's audits of the LA Sheriff's Department found pressure to change new restrictions on canine deployment from canine handlers:

> Much of the pressure to ease deployment restrictions came from the handlers themselves, and, as a result, they regained authority in April 1999 to search again for suspects wanted for auto theft, albeit with restrictions that had not existed previously to reduce the risk that juveniles would be bit.[522]

In this instance, the canine handlers enjoyed special influence because of their special expertise on the subject.

Example #2: The Erosion of Supervision

One alarming practice reported in this book is the tendency to allow the level of supervision on the street to erode by not providing the proper number of sergeants. There is a general standard that the ratio of officers per sergeant should be 7 to 1 or 8 to 1. (This ratio is also referred to as the "span of control.") Yet, the LASD Special Counsel found that in the troubled Century Station, where there had been a high number of officer-involved shootings, the ratio had at times risen as high as 20 or 25 to 1. This violated the department's own standard of 8 to 1. (See the discussions in both Chapters One and Six.)

In Philadelphia, the Integrity and Accountability Office found that the department was assigning the same number of sergeants and lieutenants to high-crime and low-crime districts. As a result, the high-crime 35th District sometimes had an officer ratio of 17 to 1, whereas the low-crime 5th District had an acceptable ratio of 8 or 7 to 1. And when some sergeants were absent because of military duty, normal vacation time, or some other factor, the ratio rose to levels of 25 or 30 to 1.[523]

It is easy to understand how this erosion of supervision can occur. A police department faces a daily struggle to remain fully staffed in the face of retirements and resignations, vacation days, sick leaves, and other routine absences. To understand this is not to accept or condone it, however. After all, police departments have been in the business of staffing patrol assignments since the days of Robert Peel. What is alarming is that police managers have not addressed this chronic problem and devised effective means of coping with it.

Another factor contributing to the erosion of supervision is the normal process of personnel reassignment. The Special Counsel's follow-up report on the Century Station is particularly instructive in this regard. A series of management changes had dramatically reduced officer-involved shootings at this troubled LASD station. But the commanders who had instituted newly intensive forms of supervision that had been instrumental in this change moved on to other assignments. In part, this was their reward for doing a good job; but it was also partly related to the normal process of personnel rotation. Their replacements, however, failed to maintain the improvements in supervision, and officer-involved shootings returned to high levels.[524] The underlying problem appears to be a failure at higher command levels to effectively orient the newly assigned commanders to the special circumstances of their new assignment.

The most disturbing aspect of routine supervision erosion is its implications for all of the other accountability mechanism, including both the traditional and the new ones described in this book. As already mentioned, patrol duty is the core police function; it is where most police–citizen encounters occur and where most misconduct occurs. If the basic supervisory function in patrol is allowed to not function properly, all of the mechanisms of accountability described in this book are also in jeopardy.

Failure to Ensure Continuity of Reform

The problems in the Los Angeles Sheriff's Department's early intervention system (the PPI) identified by the LASD's Special Counsel represent a failure to ensure the continuity of established and nationally recognized reforms. Particularly disturbing in the Special Counsel's report was the fact that some commanders were not familiar with the PPI and its capabilities. Evidently, leaders at the highest level of the department failed to ensure that all mid-level commanders, including those recently promoted, were given a complete orientation to the department's accountability programs.

Summary

How then to ensure that reforms are implemented and that departments conform to both national professional standards and their own standards? The new police accountability addresses this issue through the police auditor. A permanent external agency that has a broad license to investigate any and every aspect of police operations and to issue public reports of its findings is the best method for ensuring that police departments keep up with national standards, maintain their own standards, and not allow important reforms to whither away from neglect. The law enforcement accreditation process is, in theory, designed to achieve these goals. But as discussed in Chapter Two, the current accreditation standards are framed in terms that are far too general. Nor does the accreditation process provide the month-by-month scrutiny of police operations and the publication of quarterly or semiannual reports provided by police auditing.

THE CHALLENGE OF COMPREHENSIVE ORGANIZATIONAL CHANGE

Some police reformers assume that changing a backward and unprofessional police department is simply a matter of taking it to court and obtaining a consent decree that orders the necessary and sweeping changes. Experience has taught us that changing a large and complex bureaucracy is not that simple. The challenge is not simply one of putting into place new policies and procedures, but also one of retraining and resocializing both rank-and-file officers and supervisors. Not only must they be trained in the use of new policies and procedures, but they also need to be resocialized into understanding their purpose.

There is a sobering historical example with respect to prisoners' rights litigation that police reformers, especially litigators and monitors, should study. The prisoners' rights movement began in the late 1960s, fueled by the heady optimism of court-directed social change. In state after state, however, lawyers and their plaintiffs learned that winning a court order was the beginning and not the end of the case. In many cases, implementation took a decade or longer, often with several additional trips back into court. Budget constraints and bureaucratic inaction were only two of the many major obstacles. This is not to say that the prisoners' rights movement was a mistake. Many of the abuses it addressed are now eliminated and condemned by official professional standards and human rights declarations around the world. The point is simply that changing a complex bureaucracy to achieve worthy goals is not a simple or speedy process.[525]

A brief examination of two police departments attempting comprehensive change as a result of a settlement agreement illustrates some of the problems that reformers can encounter in this process.

The Struggle for Change in the Washington, DC, Police

The settlement of the Department of Justice investigation of the Metropolitan Police Department in Washington, DC, directed the department to develop a new use of force reporting system. Almost 2 years after the settlement agreement was signed, however, the court-appointed monitor found that in January and February 2003, officers completed the new Use of Force Incident Reports (UFIRs) in only about 25% of all incidents in which they were required. The completion rate rose to 86% in March, but then fell back to 28% in April.[526]

What accounted for this dismal record? The monitor found that there was "some confusion" among both officers and supervisors about when and how to complete a UFIR.[527] Evidently, there was a failure in the training of both officers and supervisors. It should be remembered that in a department that did not have an adequate reporting system prior to the settlement agreement, there was no established practice of reporting and review. That is to say, there was no culture of accountability. Training, therefore, involves more than just acquainting officers with a new report form they must complete, but rather socializing them into an entire process they are not familiar with—and which they are likely to resist if only because it appears as bureaucratic paperwork. Officers will also be inclined to resist because they accurately perceive that the reporting system is a tool for scrutinizing their performance far more closely than it has been in the past. In short, instituting a new system for controlling use of force involves changing the routine habits of officers, along with the values and expectations that underlie them.

The Challenge of Change in Oakland

In Oakland, California, the *Second Semi-Annual Report* of the court-appointed monitor, issued on February 18, 2004, found that although "extensive work" had been done to implement the consent decree, there were many areas of "concern." Among 50 separate tasks specified in the settlement agreement, the monitor found full compliance in only 4; the department was also "progressing without concern" in another 12. This represents only 33% of all the tasks. The department was "progressing with concern" (defined as behind schedule) in 22 tasks, it was "not in compliance" in 3, and "not in full compliance" in 9 others. By any standard, this is a decidedly mixed record of progress.[528]

Part of the problem is the sheer number of changes required by the settlement agreement. To implement them the department created a process of monthly meetings in which proposed new policies would be reviewed and ultimately approved. Within the first 6 months, it became obvious that there was not enough time for a full discussion of each policy. As policies were held over for further discussion, a backlog developed and the department began missing settlement agreement deadlines. To remedy this problem, the process was revised to include expedited review of some policies.

The monitor's reports also highlight the extent to which different tasks are interrelated, with the completion of one dependent on the development of policies or procedures specified by another task. The monitor, for example, found the department not in compliance with Task 20, which requires the department to maintain a ratio of one sergeant for eight officers. As discussed several times in this book, the failure to maintain a proper supervisory span of control represents a serious breach of professional standards and has been found in several other departments. To reach compliance, the monitor recommended, among other things, that the department realign its squads, transfer some technicians, promote additional sergeants, and conduct a department-wide review of the assignment of sergeants to see if some could be transferred to patrol.[529] In short, complying with one seemingly limited requirement calls for a sweeping reevaluation of personnel utilization in the department. This is a major undertaking in the best of circumstances, and particularly burdensome in a department faced with 49 other required tasks involving sweeping changes.

Task 01, meanwhile, requires a complete revision of the department's citizen complaint investigations manual and the training of all personnel who receive the new policies and procedures. The court-appointed monitor, however, noted that this task is linked with development of a discipline matrix, which is designed to ensure consistency of discipline (Task 45). The concept of a discipline matrix is relatively new, however, and developing one involves giving careful thought to a number of very complex issues about disciplinary procedures (e.g., how rigid should the matrix be? How much flexibility is appropriate? Should certain individual circumstances be allowed to mitigate punishment?).[530]

Task 34 requires Oakland officers "to complete a basic report on every vehicle stop, field, investigation and every detention." The monitor, however, found a compliance rate of only 26%. Interviews with officers elicited a number of explanations for this unacceptable compliance rate. Among other things, they told the monitoring team that the "Crime Reduction Teams did not receive the initial training," that "filling out the forms is too time consuming," and that the "forms are often not available." The officers' comments are highly instructive. They reflect the traditional police organizational culture that is disinterested in accountability. It is a culture that fails to appreciate the value of controls over use of force and other critical incidents and dismisses reports as "too time consuming." It is the culture of a department that fails to undertake the necessary training for an important task and cannot ensure that important report forms are available where they are needed. These are the nitty-gritty aspects of organizational culture that need to change to achieve meaningful accountability.

GETTING IT RIGHT: PROACTIVE
SELF-ASSESSMENT AND PUBLIC DISCLOSURE

In the end, the responsibility for police accountability lies with the police themselves. Status as a profession requires a commitment to professional self-regulation. External oversight is both an inherent feature of a democratic society and a practical necessity with respect to the police. Ultimately, however, police departments have to adopt and maintain continuous internal accountability programs.

The good news is that some police departments have taken notable steps in that direction. Getting it right with respect to accountability involves two steps. First, a department has to undertake a rigorous and proactive self-assessment of its own accountability policies related to the use of force, citizen complaints, race discrimination, and other issues. Second, it needs to engage community representatives in that process and fully and publicly disclose what it is doing. These are risky steps. Self-scrutiny and public disclosure may well reveal embarrassing facts. Virtually all organizations, not just police departments, traditionally avoid such potential embarrassments.

The following examples show what two police departments are doing to get it right with respect to accountability.

The Fresno Police Department: Use of Force

By the late 1990s, the Fresno, California, Police Department had a bad national reputation because of its SWAT team. There were repeated allegations of excessive use of force and a generally oppressive approach to the racial minority community. The Fresno Violent Crime Suppression Unit consisted of 30 officers who patrolled high-crime neighborhoods day and night, serving warrants on suspected drug dealers and criminals, stopping vehicles, interrogating gang members, and generally making a highly visible presence in these areas. One account described how the officers wear "subdued gray and black urban camouflage and body armor, and have at the ready, ballistic shields and helmets, M17 gas masks and rappelling gear." Additional military-style equipment was readily available in a mobile command SWAT bus. The combination of the unit's uniforms, equipment, and aggressive tactics fed all the worst fears among community activists about the militarization of the American police.[531]

After 4 years of controversy and bad publicity, the Fresno Police Department embarked on a very different course of action. It disbanded the militaristic unit in December 2001 and the chief of police created a Use of Force Subcommittee under the chief's Advisory Committees. It consisted of two captains, four lieutenants, and a staff assistant; sought input from community representatives; and had a mandate to review all aspects of use of force.

The subcommittee's final report was a thorough and extremely candid review of the department's shortcomings regarding use of force. It admitted that, "Until recently, the Department had no method to determine the number of times officers used non-lethal means to resolve potentially lethal situations, losing critical information needed to illustrate this important fact." The report found that over a 5-year period there was "a minimum of twelve cases where suspects were shot in a vehicle, and three incidents where suspects were shot after fleeing from their vehicle." It recommended additional training and review of tactics with regard to vehicle stops, pursuits, and tactics for removing people from vehicles.[532]

Perhaps the most notable aspect of the report was the statement that whereas "many other departments did not track this data for fear of providing plaintiff's attorneys with information that would assist them in suing the departments," it felt that "the ability to know what is going on outweighs the disadvantages of giving any information to plaintiff's attorneys." This argument is a startling reversal of the traditional attitude of police departments toward the collection of potentially damaging information about its own operations. Police departments have always feared that if they collect data on shootings or other forms of officer misconduct, plaintiffs' attorneys would subpoena it and use it in litigation against them. The Fresno committee's position is at the heart of the new police accountability: If a department seeks to control misconduct, it needs systematic data on its officers' performance. The collection and analysis of this kind of data is the core of early intervention systems. To be sure, there are risks related to potential litigation, but as the Fresno committee concluded, the long-term benefits far outweigh the risks.

Although the committee did not recommend any changes to the substance of the department's use of force policy, it did recommend some significant tightening of on-the-street supervision, seeking to "make sure that every effort is made to ensure that field supervisors are available and have access to officers in the field." In addition, it recommended that the Critical Incident Review Committee review shootings and other extraordinary incidents "so as to provide direction in the area of training or procedures that may need to be addressed." It also recommended changes in the existing Unusual Occurrence Report to provide additional information that would then be entered into a database in which it could be effectively analyzed. Finally, it recommended the creation of an early warning system that incorporates existing data and reports. These recommendations highlight the point argued in Chapter Three of this book that merely having a written policy on the use of force is not sufficient; effective implementation of that policy requires a number of additional administrative procedures.

A new set of accountability measures was instituted as a result of the report. Beginning in March 2003, the department's Professional Standards Unit began reviewing police reports and other force data to identify patterns and determine if policy changes are needed.[533] The quarterly report issued in late 2003 included data on use of force by Fresno officers correlated with calls for service, race of the suspects involved, the types of force used, types of incident, days of week, hours of day, police district, and suspects' actions. The data are presented in clear, easy to read graphs and charts that most people can readily comprehend.

Not only was the original subcommittee report posted on the department's official Web site, but the first in a series of follow-up reports on use of force by the department was also posted. This full disclosure about one of the most sensitive issues in policing represents the true spirit of the new police accountability.

The Seattle Police Department: Force and Racial Profiling

The Seattle Police Department has undertaken a series of steps to address the issues of use of force and racial profiling. This effort is centered in the Office of Professional Accountability (OPA), which, as described in Chapter Six, functions as a form of police auditor.

On the issue of use of force, the department issued a 20-page report on use of force by its officers, created a task force to study less than lethal force options that issued another detailed report, published an update on implementation of that report's recommendations a year later, and also published a separate report on the adoption of tasers as a less than lethal force option. All of these reports are posted on the department's Web site.[534]

The 1-year update on less than lethal weapons provided a candid look into not only the progress the department had made but also its thinking about the issue of force. It reported that the department had exceeded its goals in training officers in the Crisis Intervention Team (CIT) program—a nationally recognized program for helping officers deal with mentally disturbed persons (see Chapter Three). But it also indicated that the department was "rethinking" its goal of providing all patrol sergeants with the standard CIT training and instead offering them training in activities they are more likely to use as supervisors.

The report also candidly discussed four issues and concerns that had arisen in the operation of the CIT program. Finally, there were equally candid discussions of issues related to both the use of tasers and beanbags as less than lethal weapons.[535]

The OPA's 2003 report on racial profiling was especially detailed and candid. As previously discussed in Chapter Four, it included a detailed review of citizen complaints related to racial bias. The OPA concluded that there was no foundation to the allegations of bias in those complaints. But instead of stopping at that point, as police departments traditionally have done (in essence, declaring themselves "not guilty"), the report goes into a deeper discussion of the department's efforts to address possible racial bias. It discussed five departmental initiatives related to the issue: data collection, video cameras in patrol cars, community outreach efforts, officer training, and policy development. In particular with regard to policy development, the report explained that the department had adopted the model policy recommended by PERF on the proper use of race or ethnicity in police work and on tactics for handling traffic stops that are likely to reduce citizen resentment of the police.[536] In its assessment of the department's response to the racial profiling issue, the OPA concluded that it had made a "great start" but that there is "more work to be done" in a number of areas.[537]

In addition, the Seattle Police Department has posted its entire Departmental Manual on the Web. Traditionally, many departments have regarded their procedural manuals as nonpublic documents. This attitude of secrecy not only denies to the public basic information about official police policies, but aggravates community relations by sending a message to people that they have no right to know how the department operates.

In 2004, the Seattle Office of Professional Accountability undertook another project that is virtually unprecedented in policing. The Director of OPA (who directs internal affairs but is not a sworn officer) made personal presentations at 42 separate roll calls, communicating directly with more than 400 officers in the department. This step is significant because of the traditional conflict between police internal affairs units and the rank and file. Officers regard internal affairs as a group of "head hunters" who are simply out to get them.[538]

The roll-call presentations involved some frank discussions about misconduct investigations. Some misunderstandings were cleared up. The officers believed that OPA investigated all complaints, no matter how trivial. This reflects the rank-and-file belief that internal affairs does not respect them and listens to whatever a citizen tells them. The OPA director was able to explain that, in fact, more than 60% of all complaints received are not the subject of a full investigation (all are officially logged and subject to a preliminary investigation, however). The officers expressed concern about the length of time investigations take, leaving them with an unresolved case hanging over them. OPA acknowledged that this is a problem and is working to expedite investigations. Officers suggested that the department develop a computerized system that would allow individual officers to determine the current status of an investigation. OPA regarded this as a "great idea" and is working toward implementing it. Several other concerns were discussed and suggestions for change considered.

The significance of the OPA "internal outreach" program is that it represents a serious effort to overcome the traditional conflict between internal affairs and the rank and file. One of the obstacles to police accountability, and part of the basis for the "code of silence," is that officers do not see themselves as part of an overall accountability effort. Their relations with internal affairs are as much an attitude of "us versus them" as is their relations with the community. The OPA program—a report of which is available on the department's Web site along with the other reports—represented a genuine dialog and give and take over matters of concern.

In the end, the many reports by the Seattle Police Department present a picture of a department that is actively responding to issues of great controversy in the community, reexamining its policies and procedures, seeking input from community leaders, searching for and adopting nationally recognized policies and programs, and assessing its progress in implementing its own recommendations. All of this process, moreover, is available on the Web for anyone and everyone to read.

TOWARD THE FUTURE

If in the long run meaningful police accountability requires police departments to take responsibility for their own affairs, the history of police accountability clearly indicates that community activists have been the driving force behind the major changes of the past four decades. As Chapter Two argued, the police professionalization movement had accomplished much by the early 1960s but had shamefully neglected to address the problems of use of force and race discrimination. Change began as community activists, usually local civil rights groups, unleashed an unrelenting campaign of protest against unjustified shootings and use of excessive force that continues to this day. The deadly force policy instituted by New York City Police Commissioner Patrick V. Murphy in 1972 was adopted not because it seemed like a good idea in the abstract but because of continuing pressure from the community. By the same token, the landmark Supreme Court cases that were instrumental in provoking a new wave of police reform were brought to the Court by public interest groups, notably the ACLU.

The Department of Justice pattern or practice suits that have defined the core elements of the new police accountability are brought under Section 14141 of the 1994 Violent Crime Control Act. That section was added to the law largely because civil rights advocates in the Congress demanded it as the price for their support of conservative provisions emphasizing greater use of imprisonment. These members of Congress, in turn, were responding to their vocal constituents in the civil rights community.

This past experience suggests that continued pressure from community activists is the key to the future of the new police accountability. As this chapter has already proposed, the strategies and tactics described in this book can serve as a reform agenda for these activists. The police auditor, meanwhile, can function as the professional voice of the community with both the authority to examine police operations with an outsider's perspective and the expertise about the details of those operations that comes with experience.

The Future of Litigation

Federal pattern or practice litigation has been instrumental in bringing together disparate reform programs into the coherent package that this book labels the new police accountability. As Chapter One pointed out, all of the basic reforms contained in the settlement agreements negotiated by the Department of Justice existed in one form or another long before the 1994 law was enacted.

Given the catalytic role of federal litigation, it is necessary to discuss its future role. To the surprise of many observers, the Bush administration Department of Justice did not stop using Section 14141. In fact, a majority of the 19 settlement agreements have been reached since John Ashcroft became

Attorney General, including the 2002 Cincinnati Memorandum of Agreement that, in many respects, goes further than other agreements (although given the process of investigating a department and developing a settlement agreement, many of these were carry-overs from the Clinton administration).

Nonetheless, it would be a mistake to assume that the U.S. Department of Justice will continue to play the leading role in pursuing police accountability. First, the Special Litigation Section of the Civil Rights Division has limited resources and can hardly be expected to address the problems in all 18,000 state and local law enforcement agencies in this country. In addition, no one can predict what changes in priorities might occur in the future, as a result of budget constraints on the Department of Justice or deliberate changes in policy.

Even a modest federal role in the future leaves open other important opportunities for accountability related litigation. First, two major settlements in recent years have resulted from private suits over police misconduct: in Philadelphia and Oakland. Nothing precludes future litigation by private parties. Second, two settlements have arisen from suits brought by state attorneys general: in California (the Riverside Police Department) and in New York (the small town police department in Walkill, New York). Future suits of this sort could be facilitated if each of the 50 state legislatures enacted legislation following the language of Section 14141 and authorizing their state attorneys general to bring pattern or practice suits. The Riverside and Walkill suits were brought under more general legal authority, but specific enabling legislation would greatly strengthen the hand of attorneys general.

In the End: A Mixed Approach to Police Accountability

Litigation is not the sole answer to the problem of police accountability. The best approach is a mix of systems. First, police departments can and should adopt the proactive efforts that Fresno and Seattle have already undertaken. Second, community activists can and should continue to demand the highest standards of accountability, including proactive efforts. Third, to ensure the continuity of reform, cities and counties can and should create their own local police auditors.

Misconduct by police officers has been a problem in the United States since the creation of the first police departments. In addition to the violations of the rights and dignity of innumerable people, it is the source of serious racial and ethnic tensions. The police professionalization movement accomplished much during the twentieth century in raising standards for American police departments. But it also fell far short of what was needed in the area of accountability. In that context, the new police accountability represents a historic development. It has the potential for establishing truly professional standards in policing. Whether this potential will be realized is far from certain. Responsibility lies with the law enforcement profession and community leaders alike to pursue the strategies and tactics of the new accountability, to implement these measures, to assess their effectiveness, and, if they are found wanting, to develop and pursue alternative measures.

NOTES

1. Merrick Bobb, Special Counsel, *15th Semiannual Report* (Los Angeles: Los Angeles County Sheriff's Department, 2002), 16. The reports of the Special Counsel are available at www.parc.info.

2. Ibid.

3. Merrick Bobb, Special Counsel, *9th Semiannual Report* (Los Angeles: Los Angeles County Sheriff's Department, 1998), 8. Report available at www.parc.info.

4. Ibid.

5. Merrick Bobb, *15th Semiannual Report*, 12.

6. *United States v. New Jersey,* Consent Decree (1999). All of the consent decrees, memoranda of agreement, and letters negotiated by the U.S. Department of Justice are available at www.usdoj.gov/crt/split.

7. This information is based on personal observations by the author at a large police department that prefers to remain anonymous.

8. Samuel Walker, *Early Intervention Systems for Law Enforcement Agencies: A Planning and Management Guide* (Washington, DC: U.S. Department of Justice, 2003). The report is available at www.cops.usdoj.gov and www.ncjrs.org, NCJ 201245.

9. A summary of the argument in this book appeared in Samuel Walker, "The New Paradigm of Police Accountability: The U.S. Justice Department 'Pattern or Practice' Suits in Context," *St. Louis University Public Law Review* XXII, no. 1 (2003): 3–52.

10. Barbara Armacost, "Organizational Culture and Police Misconduct," *George Washington Law Review* 72 (March 2004): 457–59.

11. Samuel Walker, *A Critical History of Police Reform* (Lexington, MA: Lexington Books, 1977).

12. Bobb, *15th Semiannual Report*, 10.

13. Armacost, "Organizational Culture and Police Misconduct," 455.

14. 42 U.S.C. §§ 14141. Cause of action.

(a) Unlawful conduct
It shall be unlawful for any governmental authority, or any agent thereof, or any person acting on behalf of a governmental authority, to engage in a pattern or practice of conduct by law enforcement officers or by officials

or employees of any governmental agency with responsibility for the administration of juvenile justice or the incarceration of juveniles that deprives persons of rights, privileges, or immunities secured or protected by the Constitution or laws of the United States.

(b) Civil action by Attorney General
Whenever the Attorney General has reasonable cause to believe that a violation of paragraph (1) has occurred, the Attorney General, for or in the name of the United States, may in a civil action obtain appropriate equitable and declaratory relief to eliminate the pattern or practice.

15. United States Department of Justice, *Principles for Promoting Police Integrity* (Washington, DC: Department of Justice, 2001). Available at www.ncjrs.org, NCJ 186189.

16. The settlements are available at www.usdoj.gov/crt/split.

17. These reports can be found on the Web sites of the various police departments, for example: Los Angeles: www.lapdonline.org.

18. Los Angeles Sheriff's Department: www.parc.info. Boise: www.boiseombudsman.org. San Jose: www.ci.san-jose.ca.us/ipa/home.html.

19. Several are available on the PARC Web site: www.parc.info.

20. The literature on American policing is disturbingly scant on the subject of accountability. This summary of the dimensions of accountability is taken from the Independent Commission on Policing in Northern Ireland (the Patten Commission) *Report* (2000) at Chap. 5, "Accountability I: The Present Position," Sec 5.4.

21. David Bayley, *Police for the Future* (New York: Oxford University Press, 1994).

22. The term *legitimacy* is increasingly used to encompass the related issues of police compliance with the law and citizen perceptions of the police. See National Academy of Sciences, *Fairness and Effectiveness in Policing: The Evidence* (Washington, DC: National Academy Press, 2004).

23. Samuel Walker and Charles M. Katz, *The Police in America: An Introduction*, 5th ed. (New York: McGraw-Hill, 2005), 473–75.

24. This point is the basic theme of Samuel Walker, *Popular Justice: A History of American Criminal Justice,* 2nd ed. (New York: Oxford University Press, 1998).

25. This is the central theme in Walker, *Popular Justice: A History of American Criminal Justice.* Jerome Skolnick, *Justice Without Trial: Law Enforcement in a Democratic Society*, 3rd ed. (New York: Macmillan, 1994).

26. Herbert Packer, *The Limits of the Criminal Sanction* (Stanford: Stanford University Press, 1968), Chap. 8, 149–73.

27. *Cincinnati Enquirer,* "2003: the Year in Review," December 31, 2003.

28. Lou Cannon, *Official Negligence: How Rodney King Changed Los Angeles and the LAPD* (New York: Times Books, 1997).

29. National Advisory Commission on Civil Disorders, *Report* (New York: Bantam Books, 1968).

30. Richard A. Leo and George C. Thomas, eds., *The Miranda Debate: Law, Justice, and Policing* (Boston: Northeastern University Press, 1998).

31. Jack Greene, "Community Policing in America: Changing the Nature, Structure, and Function of the Police," in *Criminal Justice 2000. V. 3: Policies, Processes, and Decisions of the Criminal Justice System,* ed. Julie Horney, (Washington, DC: U.S. Department of Justice, 2000), 299. www.ncjrs.org, NCJ 182410. Michael S. Scott, *Problem-Oriented Policing: Reflections on the First 20 Years* (Washington, DC: Department of Justice, 2000). Available at www.cops.usdoj.gov.

32. Walker and Katz, *The Police in America,* 4th ed., 387–89.

33. David L. Carter, Allen D. Sapp, and Darrel W. Stephens, *The State of Police Education: Policy Direction for the 21st Century* (Washington, DC: Police Executive Research Forum, 1989). Trends in policing are reviewed in Walker and Katz, *The Police in America,* 431–38.

34. The best summary of these best practices is the Department of Justice, *Principles for Promoting Police Integrity.* This report was developed through a series of Department of Justice sponsored conferences as workshops in the preceding years. See U.S. Department of Justice, "Attorney General's Conference: Strengthening Police-Community Relationships," *Summary Report.* Washington, DC, June 1999.

35. Walker, "The New Paradigm of Police Accountability: The U.S. Justice Department 'Pattern or Practice' Suits in Context." (see n. 9)

36. Herman Goldstein, personal communication with author, June 2003.

37. The consent decrees and memoranda of understanding negotiated by the Justice Department are available at www.usdoj.gov/crt/split.

38. Debra Livingston, "Police Reform and the Department of Justice: An Essay on Accountability," *Buffalo Criminal Law Review* 2 (1999): 848.

39. Erwin Chemerinsky, *An Independent Analysis of the Los Angeles Police Department's Board of Inquiry Report on the Rampart Scandal* (Los Angeles: Police Protective League, 2000). The classic work on the norms of secrecy in the police subculture is William A. Westley, *Violence and the Police* (Cambridge: MIT Press, 1970).

40. National Institute of Medicine, *To Err is Human* (Washington, DC: National Academy Press, 1999).

41. See for example the City of Baltimore Citistats program at www.ci.baltimore.md.us.

42. James J. Willis, Stephen D. Mastrofski, David Weisburd, and Rosann Greenspan, *Compstat and Organizational Change in the Lowell Police Department: Challenges and Opportunities* (Washington, DC: The Police Foundation, 2004).

43. Michael S. Scott, *Problem-Oriented Policing: Reflections on the First 20 Years* (Washington, DC: Department of Justice, 2002). Available at www.ncjrs.org.

44. Bobb, Special Counsel, *15th Semiannual Report,* 34.

45. Summarized in Walker, "The New Paradigm of Police Accountability."

46. Commission to Investigate Allegations of Police Corruption and the Anti-Corruptions Procedures of the Police Department [Mollen Commission], *Commission Report* (New York, 1994). Available at www.parc.info.

47. Department of Justice, *Principles for Promoting Police Integrity.*

48. The reports of the OIR and the Special Counsel are available at www .parc.info.

49. Samuel Walker, *A Critical History of Police Reform* (Lexington, MA: Lexington Books, 1977).

50. Barbara Armacost, "Organizational Culture and Police Misconduct," *George Washington Law Review* 72 (March 2004): 455.

51. Walker, *A Critical History of Police Reform.*

52. See the provocative discussion of the development of a police monopoly over their professional mandate in Peter K. Manning, *Police Work* (Cambridge: MIT Press, 1977). A critique of this insular professional monopoly over the delivery of public services is one of the core principles of the community policing movement. George L. Kelling and Mark H. Moore, *The Evolving Strategy of Policing, Perspectives on Policing,* no. 4 (Washington, DC: U.S. Justice Department, 1988).

53. For a contemporary account of the fierce reaction to the Supreme Court's decisions on the police, see Fred P. Graham, *The Self-Inflicted Wound* (New York: Macmillan, 1970). On the reaction to police unions, see Peter Feuille, *Police Unionism* (Lexington, MA: Lexington Books, 1977). See the various contributions in the valuable collection, William A. Geller, ed., *Police Leadership in America: Crisis and Opportunity* (New York: Praeger, 1985).

54. Walker, *A Critical History of Police Reform.*

55. It is possible to benchmark improvements in policing by comparing the data on police in The Cleveland Foundation, *Cleveland Survey* (Cleveland: Cleveland Foundation, 1922) (the first of the modern crime commissions), National Commission on Law Observance and Enforcement, *The Police* (Washington, DC: Government Printing Office, 1931) (the first national crime commission), and the President's Commission on Law Enforcement, *Task Force Report: The Police* (Washington, DC: Government Printing Office, 1967).

56. The characterization of the police as "adjuncts to the machine" is in Robert Fogelson, *Big City Police* (Cambridge: MIT Press, 1977), 13.

57. National Advisory Commission on Civil Disorders, *Report.* (see n. 29)

58. Samuel Walker, *Police Accountability: The Role of Citizen Oversight* (Belmont, CA: Wadsworth, 2001), 180–83, 193–201.

59. National Advisory Commission on Civil Disorders, *Report,* 301.

60. O. W. Wilson and Roy C. McLaren, *Police Administration,* 4th ed. (New York: McGraw-Hill, 1974), 136–41.

61. Samuel Walker, "The Creation of the Contemporary Criminal Justice Paradigm: The American Bar Foundation Survey of Criminal Justice, 1956-1969," *Justice Quarterly,* 9 (1992): 201.

62. National Academy of Sciences, *Fairness and Effectiveness in Policing: The Evidence* (Washington, DC: National Academy Press, 2004), 34.

63. Philadelphia Police Study Task Force, *Philadelphia and its Police* (Philadelphia, 1987), 48–49.

64. City of Buffalo, Press Release, July 17, 2003, www.ci.buffalo.ny.us.

65. Merrick Bobb, Special Counsel, *9th Semiannual Report* (Los Angeles: Los Angeles Sheriff's Department, 1998). Available at www.parc.info. *People of California v. City of Riverside,* Stipulated Judgment (March 2001), Para 58. Available at www.ci .riverside.ca.us/rpd. *NAACP, ACLU, and the Barrio Project v. City of Philadelphia,* Stipulated Agreement, September, 1996.

66. Robin Sheppard Engel, *How Police Supervisory Styles Influence Patrol Officer Behavior* (Washington, DC: Department of Justice, 2003). Available at www.ncjrs.org, NCJ 194078.

67. Los Angeles Police Department, *Board of Inquiry Report on the Rampart Incident* (Los Angeles: Los Angeles Police Department, 2000), 335.

68. Mollen Commission, *Report* (New York, 1994), 81. Available at www .parc.info.

69. Timothy Oettmeier and Mary Ann Wycoff, *Personnel Performance Evaluations in the Community Policing Context* (Washington, DC: Police Executive Research Forum, 1997), 5.

70. Frank Landy, *Performance Appraisal in Police Departments* (Washington, DC: The Police Foundation, 1977).

71. *Detroit Free Press*, "City Had a Bad Cop Warning," Dec 29, 2000, 1.

72. Carol Archbold, *Police Accountability, Risk Management, and Legal Advising* (New York: LFB Scholarly Publishing, 2004).

73. President's Commission on Law Enforcement and Administration of Justice, *Task Force Report: The Police* (Washington, DC: Government Printing Office, 1967), 63–67. American Bar Association, *Standards Relating to the Urban Police Function,* 2nd ed. (Boston: Little, Brown, 1980), Standard 1-7.9.

74. Philadelphia, Integrity and Accountability Office, *Disciplinary System* (Philadelphia: Philadelphia Police Department, 2001), 38.

75. See, for example, the shocking report by the National Academy of Sciences on accidental deaths in American hospitals: Institute of Medicine, *To Err Is Human: Building a Safer Health System* (Washington, DC: National Academy Press, 2000).

76. Commission on Accreditation for Law Enforcement Agencies, *Standards for Law Enforcement Agencies,* 4th ed. (Fairfax, VA: CALEA, 1998), Standard 51.1, 53.2.

77. The CALEA Web site is www.calea.org.

78. Commission on Accreditation for Law Enforcement Agencies, *Standards for Law Enforcement Agencies,* Standard 41.2.2.

79. Ibid., Standard 35.1.1.

80. American Correctional Association, *Performance-Based Standards for Adult Community Residential Services*, 4th ed. (Lanham, MD: American Correctional Association, 2000).

81. See the commentary on the failure of the law enforcement profession and state governments in this regard in American Bar Association, *Standards Relating to the Urban Police Function,* 2nd ed. (Boston: Little, Brown, 1980), particularly Standard 1-5.3.

82. On the development of a pervasive "rights culture" in America, see Samuel Walker, *The Rights Revolution* (New York: Oxford University Press, 1994).

83. On the structure of American policing and its impact, see National Academy of Sciences, *Fairness and Effectiveness in Policing: The Evidence* (Washington, DC: National Academy Press, 2004), chap. 3.

84. Malcolm M. Feeley and Edward L. Rubin, *Judicial Policy Making and the Modern State* (New York: Cambridge University Press, 1998).

85. Richard Leo and George C. Thomas, III, eds., *The Miranda Debate: Law, Justice, and Policing* (Boston: Northeastern University Press, 1998).

86. Myron Orfield, "The Exclusionary Rule and Deterrence: An Empirical Study of Chicago Narcotics Officers," *University of Chicago Law Review* 54 (Summer 1983): 1016–55. This study has one of the most detailed and illuminating accounts of the changes in training and relationships between the police and local prosecutors as a result of the *Mapp* decision.

87. Samuel Walker, *Taming the System: The Control of Discretion in Criminal Justice, 1950–1990* (New York: Oxford University Press, 1993).

88. On the prisoner's rights movement: Malcolm M. Feeley and Edward L. Rubin, *Judicial Policy Making and the Modern State* (New York: Cambridge University Press, 1998).

89. Samuel Walker, "Historical Roots of the Legal Control of Police Behavior," in *Police Innovation and the Control of the Police,* ed. David Weisburd and Craig Uchida, 32–55 (New York: Springer-Verlag, 1993).

90. Anthony M. Amsterdam, "Perspectives on the Fourth Amendment," *Minnesota Law Review* 58 (1973–74): 349–577.American Bar Association, *Standards Relating to the Urban Police Function*, Standard 1-5.3, "Sanctions," and accompanying commentary.

91. Dallin H. Oaks, "Studying the Exclusionary Rule in Search and Seizure," *University of Chicago Law Review* 37 (Summer 1970): 665–757.

92. See the research by Richard Leo, excerpted in Leo and Thomas, eds., *The Miranda Debate.*

93. American Bar Association, *Standards Relating to the Urban Police Function.*

94. Mary M. Cheh, "Are Law Suits an Answer to Police Brutality?" in *And Justice for All: Understanding and Controlling Police Abuse of Force,* ed. William A. Geller and Hans Toch, 256–8 (Washington, DC: Police Executive Research Forum, 1995).

95. "Project: Suing the Police in Federal Court," *Yale Law Journal* 88 (1979): 781. Candace McCoy, "Lawsuits Against Police: What Impact Do They Really Have," *Criminal Law Bulletin* 20 (1984): 53. Human Rights Watch concluded that civil litigation "must always be available, but cannot be a substitute for police departmental mechanisms of accountability or prosecutorial action." *Shielded from Justice* (Washington, DC: Human Rights Watch, 1998), 85.

96. The 1992 Kolts investigation of the Los Angeles Sheriff's Department was prompted primarily by concern about civil litigation costs. James G. Kolts, *The Los Angeles Sheriff's Department* (Los Angeles: LASD, 1992). Available at www.parc.info.

97. Human Rights Watch, *Shielded from Justice,* 81.

98. Armacost, "Organizational Culture and Police Misconduct," 474–5.

99. Carol Archbold, *Police Accountability, Risk Management, and Legal Advising* (New York: LFB Scholarly Publishing, 2004). Note also that the leading textbook on police administration contains no reference to risk management, James J. Fyfe, Jack R. Greene, and others, *Police Administration,* 5th ed. (New York: McGraw-Hill, 1997) This is the updated version of the classic O. W. Wilson and Roy C. McLlaren, *Police Administration,* 4th ed. (New York: McGraw-Hill, 1977).

100. Kolts, *The Los Angeles Sheriff's Department.*

101. The work of the Special Litigation Section of the Civil Rights Division is available at www.usdoj.gov/crt/split.

102. Feeley and Rubin, *Judicial Policy Making and the Modern State.*

103. *Rizzo v. Goode,* 423 U.S. 362 (1976). This discussion is based on several conversations with Professor Goldstein over the years, and I am much in debt to him for his insights.

104. Feeley and Rubin, *Judicial Policy Making and the Modern State,* especially the reference to *Rizzo v. Goode* on pp. 250–251.

105. Vera Institute of Justice, *Prosecuting Police Misconduct* (New York: Vera Institute of Justice, 1998). Available at www.vera.org. "Securing Police Compliance with Constitutional Limitations: The Exclusionary Rule and other Devices," in National Commission on the Causes and Prevention of Violence, *Report* (New York: Bantam Books, 1970), 405–7. Human Rights Watch, *Shielded from Justice,* 85–103. "Criminal prosecutions and other kinds of law suits have not played a major role in addressing the problem of excessive force by the police . . . ." Cheh, "Are Law Suits an Answer to Police Brutality?" 234.

106. The phrase is from Lawrence W. Sherman's study of police corruption. Lawrence W. Sherman, *Scandal and Reform* (Berkeley: University of California Press, 1978).

107. A useful collection and analysis is Anthony M. Platt, ed., *The Politics of Riot Commissions, 1917–1970* (New York: Collier Books, 1971).

108. Christopher Commission, *Report of the Independent Commission on the Los Angeles Police Department* (Los Angeles, 1991). Available at www.parc.info. Human Rights Watch, *Shielded from Justice,* 44–46.

109. National Commission on Law Observance and Enforcement, *Lawlessness in Law Enforcement* (Washington, DC: Government Printing Office, 1931).

110. President's Commission on Law Enforcement and Administration of Justice, *The Challenge of Crime in a Free Society* (Washington, DC: Government Printing Office, 1967). See also the accompanying *Task Force Report: The Police* (Washington, DC: Government Printing Office, 1967).

111. American Bar Association, *Standards Relating to the Urban Police Function,* 2nd ed. (1980).

112. Samuel Walker, "Setting the Standards: The Efforts and Impact of Blue-Ribbon Commissions on the Police," in *Police Leadership in America: Crisis and Opportunity,* ed. William A. Geller (New York: Praeger, 1985, 354–70).

113. Events in Los Angeles in the decade of the 1990s offer one notable example of this process. The original 1991 beating of Rodney King led to the Christopher Commission. It prompted two follow-up reports assessing implementation of the recommendations. A few years later the Rampart scandal erupted, prompting three separate reports and the Department of Justice investigation that resulted in the current consent decree. The notable exception to this rule, also in Los Angeles ironically, is the sequence of events surrounding the Sheriff's Department, beginning with the Kolts Commission: James G. Kolts, *The Los Angeles County Sheriff's Department*, which in turn led to the creation of the permanent office of the Special Counsel. See the discussion in Chapter Six of this book.

114. Walker, *Police Accountability: The Role of Citizen Oversight*.

115. Ibid.

116. The brief history and quick demise of the New York City CCRB is told in Algernon Black, *The People and the Police* (New York: McGraw-Hill, 1968).

117. Walker, *Police Accountability: The Role of Citizen Oversight*.

118. New York Civil Liberties Union, *Five Years of Civilian Review: A Mandate Unfulfilled* (New York: New York Civil Liberties Union, 1998).

119. New Orleans, Police Civilian-Review Task Force, *Draft Report* (2002), 16. Available at www.new-orleans.la.us/.

120. Merrick Bobb and Julio A. Thompson, *The Detroit Police Department* (1997). Unpublished report, posted on the Web by the *Detroit Free Press*: www.freep.com.

121. A discussion of the different models of oversight agencies, and the meaning of "independence" in this context is in Walker, *Police Accountability: The Role of Citizen Oversight*, 61–63. See also Richard Terrill, *Alternative Perceptions of Independence in Civilian Oversight, Journal of Police Science and Administration* 17 (1990): 77–83.

122. Comparative data on staffing levels is in New York Civil Liberties Union, *Civilian Review Agencies: A Comparative Study* (New York: New York Civil Liberties Union, 1993). See in particular the low level of staffing for the Cincinnati Office of Municipal Investigations, with only one investigator for about 1,000 sworn officers.

123. A visit to the New Orleans Office of Municipal Investigations in 1995 by a Human Rights Watch investigator found that "the office was absolutely silent, no phones were ringing, and some staffers were playing computer video games." Human Rights Watch, *Shielded from Justice,* 259.

124. See the litigation sponsored by the Fraternal Order of Police, reported in Philadelphia, Police Advisory Commission, *Annual Report* (Philadelphia: Police Advisory Commission, 1997), 2–3. Available at www.phila.gov/pac.

125. The first serious discussion of this issue was Walter Gellhorn, *When Citizens Complain* (New York: Columbia University Press, 1966), 191. See also Walker, *Police Accountability: The Role of Citizen Oversight,* 121–37.

126. Walker, *Police Accountability: The Role of Citizen Oversight,* 137–8.

127. National Academy of Sciences, *Fairness and Effectiveness in Policing: The Evidence* (Washington, DC: National Academy Press, 2004).

128. A recent and notable exception to this rule is Robin Sheppard Engel, *How Police Supervisory Styles Influence Patrol Officer Behavior* (Washington, DC: Department of Justice, 2003). Available at www.ncjrs.org, NCJ 194078. The scant body of literature she is able to cite is eloquent testimony to the neglect of this critical subject.

129. Walker, "Setting the Standards: The Efforts and Impact of Blue-Ribbon Commissions on the Police," 354.

130. Department of Justice, *Principles for Promoting Police Integrity* (Washington, DC: U.S. Department of Justice, 2001), 4. Available at www.ncjrs.org, NCJ 186189.

131. James J. Fyfe, "Police Use of Deadly Force: Research and Reform," *Justice Quarterly* 5 (June 1988): 168–9.

132. O. W. Wilson, *Police Administration*, 2nd ed. (New York: McGraw-Hill, 1963).

133. Philadelphia Police Department, Integrity and Accountability Office, *Use of Force* (Philadelphia: Philadelphia Police Department, July 1999), 10. On the role of the Integrity and Accountability Office as a form of external citizen oversight, see Chap. 6.

134. Fyfe, "Police Use of Deadly Force." The University of Wisconsin Law School Library has a valuable collection of old police department manuals.

135. Los Angeles Police Department, *Report of the Independent Monitor for the Los Angeles Police Department, Report for the Quarter Ending December 31, 2003* (Los Angeles, 2004), 4. All the monitor's reports are available at www.lapdonline.com.

136. National Advisory Commission on Civil Disorders, *Report* (New York: Bantam Books, 1968).

137. Bureau of Justice Statistics, *Policing and Homicide, 1976–98: Justifiable Homicide by Police, Police Officers Murdered by Felons* (Washington, DC: Department of Justice, 2001). www.ncjrs.org, NCJ 180987.

138. Jerry R. Sparger and David J. Giacopassi, "Memphis Revisited: A Reexamination of Police Shootings After the *Garner* Decision," *Justice Quarterly* 9 (June 1992): 211–25.

139. Astonishingly, Murphy does not mention this historic achievement in his memoirs: Patrick V. Murphy and Thomas Plate, *Commissioner* (New York: Simon & Schuster, 1977).

140. James J. Fyfe, "Administrative Intervention on Police Shooting Discretion: An Empirical Examination," *Journal of Criminal Justice* 9 (Winter 1979): 309–23.

141. Fyfe, "Administrative Interventions." The most comprehensive review of the deadly force policy is William A. Geller and Michael S. Scott, *Deadly Force: What We Know* (Washington, DC: Police Executive Research Forum, 1992).

142. Catherine H. Milton and others, *Police Use of Deadly Force* (Washington, DC: The Police Foundation, 1977), 138.

143. U.S. Commission on Civil Rights, *Who is Guarding the Guardians? A Report on Police Practices* (Washington, DC: Government Printing Office, 1981), Finding 3.1, 156.

144. A 2001 report on use of force by the Seattle Police Department reported that "The national standard among police agencies is not to fire warning shots." The report also noted that the prohibition on shots to wound was now a national standard and commented that "Movies and television programs make it seem that shooting at a person's arm or leg is easily done. In real life, such a shot is both improbable and risky." Seattle Police Department, *Use of Force by Seattle Police Department Officers* (November 2001), 12. Available at www.cityofseattle.org/police.

145. Bureau of Justice Statistics, *Policing and Homicide, 1976–98.*

146. Sparger and Giacopassi, "Memphis Revisited: A Reexamination of Police Shootings after the *Garner* Decision."

147. Compare, for example, the 1997 consent decree with the Pittsburgh Police Department with the far longer and more elaborate 2001 consent decree covering the Los Angeles Police Department. Available at www.usdoj.gov/crt/split.

148. Police Assessment Resource Center, *Portland Police Bureau, Officer-Involved Shootings and In-Custody Deaths* (Los Angeles: PARC, 2003). Available at www.parc.info.

149. U.S. Department of Justice, *Investigation of the Schenectady Police Department, Letter to Michael T. Brockbanck, Schenectady Corporation Counsel,* March 19, 2003, www.usdoj.gov/crt/split.

150. Kenneth C. Davis, *Police Discretion* (St. Paul, MN: West, 1975). Kenneth C. Davis, *Discretionary Justice: A Preliminary Inquiry* (Urbana: University of Illinois Press, 1971).

151. Samuel Walker, *Taming the System: The Control of Discretion in Criminal Justice, 1950–1990* (New York: Oxford University Press, 1993).

152. Walker, *Taming the System.*

153. Davis, *Police Discretion,* 140.

154. The history of police deadly force policies is in Geller and Scott, *Deadly Force: What We Know.*

155. Davis, *Police Discretion,* 145.

156. Geoffrey P. Alpert and Roger G. Dunham, *Police Pursuit Driving: Controlling Responses to Emergency Situations* (New York: Greenwood Press, 1990).

157. Substantial research has found that officers made arrest decisions on the basis of nonlegal factors such as the preference of the victim or the nature of the relationship between the victim and the offender. Donald Black, *The Manners and Customs of the Police* (New York: Academic Press, 1980), 85.

158. Department of Justice, *Principles for Promoting Police Integrity,* 3: "Law enforcement agencies must recognize and respect the value and dignity of every person. In vesting law enforcement officers with the lawful authority to use force to protect the public welfare, a careful balancing of all human interests is required."

159. Kansas City Police Department, *Procedural Instruction 01-3,* "Use of Force" (3/14/01). Available at www.kcpd.org.

160. California Peace Officers Association, *Sample Policy,* "Use of Force." www.cpoa.org.

161. Philadelphia, Integrity and Accountability Office, *Use of Force,* 23.

162. Robert Stewart, Consultant, *Report on the Louisville Police Department* (Louisville, KY: Louisville Police Department, 2002). Copy in author's files.

163. Bureau of Justice Statistics, *Use of Force by Police: Overview of National and Local Data* (Washington, DC: U.S. Department of Justice, 1999). Available at www.ncjrs.org, NCJ 176330.

164. Department of Justice, *Principles for Promoting Police Integrity*, 5–6.

165. Ibid., 5–6.

166. The Kansas City department policy states that: "Officers are *not* [emphasis in original] to strike anyone in the head with a weapon (e.g., baton, shotgun, handgun, etc.) in order to gain or maintain control or compliance." Kansas City Police Department, *Procedural Instruction C 01-3*, "Use of Force." Available at www.kcpd .org.

167. Omaha Police Department, *Standard Operating Procedure Manual*, "Use of Deadly Force and Non-Deadly Force" (December 1992).

168. San Diego Police Department, *Use of Force Task Force Final Report* (August 2001), Attachment to report. Available at www.sandiego.gov/police.

169. Samuel Walker, *Early Intervention Systems for Law Enforcement Agencies: A Management and Planning Guide* (Washington, DC: Department of Justice, 2003). Available at www.ncjrs.org, NCJ 201245.

170. "We recommend that the SPD adopt a policy that requires reporting for all uses of physical or instrumental force beyond unresisted handcuffing on a form dedicated solely to recording use of force information." U.S. Department of Justice *Investigation of the Schenectady Police Department, Letter to Michael T. Brockbanck, Schenectady Corporation Counsel* (March 19, 2003), 11. Available at www.usdoj .gov/crt/split.

171. *United States v. City of Detroit, Consent Judgment* (June 12, 2003). Available at www.usdoj.gov/crt/split.

172. U.S. Department of Justice, *Investigation of the Schenectady Police Department, Letter to Michael T. Brockbanck, Schenectady Corporation Counsel.*

173. U. S. Department of Justice, *Investigation of the Miami Police Department, Letter to Alejandro Vilarello, City Attorney,* March 13, 2003. Available at www.usdoj.gov/crt/split.

174. The accreditation process requires a comprehensive review of a department's policies and procedures, but accreditation is a voluntary process, and by 2004, only about 500 law enforcement agencies in the U.S. were accredited. On the accreditation process, see www.calea.org.

175. California Penal Code Section 835a provides that: "Any peace officer who has reasonable cause to believe that the person to be arrested has committed a public offense may use reasonable force to effect the arrest, to prevent escape or to overcome resistance."

176. Jerome H. Skolnick and James J. Fyfe, *Above the Law: Police Abuse and the Excessive Use of Force* (New York: Free Press, 1993).

177. Department of Justice, *United States v. City of Detroit, Consent Judgment* (June 12, 2003). Available at www.usdoj.gov/crt/split.

178. *United States v. City of Buffalo, Memorandum of Agreement* (September 19, 2002). Available at www.usdoj.gov/crt/split.

179. Kenneth Adams, "What We Know About Police Use of Force," in Bureau of Justice Statistics, *Police Use of Force* (Washington, DC: U.S. Department of Justice, 1999) 11.

180. National Academy of Science, *Fairness and Effectiveness in Policing: The Evidence* (Washington, DC: National Academy Press, 2004), 284–85.

181. Department of Justice, *United States v. City of Detroit,* Consent Judgment (June 12, 2003).

182. Las Vegas Metropolitan Police Department, *Procedural Order PO-03-04,* "Use of Force." (Las Vegas: Las Vegas Metropolitan Police Department, 2004). Copy in possession of author.

183. U.S. Department of Justice, *Investigation of the Schenectady Police Department,* 9.

184. Joel H. Garner and Christopher D. Maxwell, "Measuring the Amount of Force Used by and Against the Police in Six Jurisdictions," Bureau of Justice Statistics, *Use of Force by Police: Overview of National and Local Data,* 37–39, www.ncjrs.org, NCJ 176330. The use of force continuum is recommended by the Department of Justice, *Principles for Promoting Police Integrity,* 4.

185. California Peace Officers' Association, Sample Policy, *Use of Force.* Available at: www.cpoa.org.

186. Justice Department, *Principles for Promoting Police Integrity,* 4.

187. Geoffrey P. Alpert, "The Force Factor: Measuring and Assessing Police Use of Force and Suspect Resistance," in Bureau of Justice Statistics, *Use of Force by Police,* 45–60.

188. Kansas City Police Department, *Procedural Instruction C 01-3,* "Use of Force." Available at www.kcpd.org.

189. United States Department of Justice and the Metropolitan Police Department of the District of Columbia, *Memorandum of Agreement* (June 13, 2001), Para 37. Available at www.usdoj.gov/crt/split.

190. Skolnick and Fyfe, *Above the Law.*

191. See, for example, the Web site of the Verbal Judo Institute: www.verbal-judo.org.

192. Peter Scharf and Arnold Binder, *The Badge and the Bullet: Police Use of Deadly Force* (New York: Praeger, 1983), Ch. 5, but especially p. 117.

193. Bureau of Justice Assistance, Practitioners Perspectives, *The Memphis, Tennessee, Police Department's Crisis Intervention Team* (2000). Available at www.ncjrs.org, NCJ 182501.

194. Police Assessment Resource Center, *Portland Police Bureau, Officer-Involved Shootings and In-Custody Deaths* (Los Angeles: PARC, 2003), 204. Available at www.parc.info.

195. Richard Jerome, *Police Oversight Project, City of Albuquerque* (Los Angeles: PARC, 2002). Seattle Police Department, *Less Lethal Options Program— Year 1* (May 2002), 16. Available at www.cityofseattle.gov/police.

196. Alpert and Dunham, *Police Pursuit Driving: Controlling Responses to Emergency Situations.*

197. Ibid.

198. Geoffrey P. Alpert, *Police Pursuit: Policies and Training* (Washington, DC: U.S. Department of Justice, 1997).

199. Merrick Bobb, Special Counsel, *16th Semiannual Report* (Los Angeles: Los Angeles Sheriff's Department, 2003), 11. Available at www.parc.info.

200. Ibid., 5.

201. Ibid., 7.

202. Police Assessment Resource Center (PARC), *Portland Police Bureau, Officer-Involved Shootings and In-Custody Deaths* (Los Angeles: Police Assessment Resource Center, 2003), 185. www.parc.info.

203. Ibid., 27.

204. Cincinnati Police Department, *Status Report of the Independent Monitor,* 7. "The Planning Section reviewed the Collingswood policy and used it as the basis for the new CPD foot pursuit policy." Available at www.cincinnati-oh.gov/police/pages/ -3039-/.

205. International Association of Chiefs of Police, *Foot Pursuits*, Concepts and Issues Paper, February 2003. Available from the IACP, www.theiacp.org.

206. Department of Justice, *Principles for Promoting Police Integrity,* 4.

207. *United States v. City of Cincinnati,* Memorandum of Agreement, (April 12, 2002), Sec. IV. C. Available at www.usdoj.gov/crt/split.

208. United States Justice Department and the Metropolitan Police Department of the District of Columbia, *Memorandum of Agreement* (June 13, 2001), Par. 45–46.

209. *United States v. City of Los Angeles* (June 15, 2001), Consent Decree, Par. 41(b).

210. Philadelphia, Integrity and Accountability Office, *Use of Force* (Philadelphia: Integrity and Accountability Office, July 1999), 10.

211. Department of Justice, *Letter to Alejandro Vilarello, City Attorney, City of Miami,* March 13, 2003. Available at www.usdoj.gov/crt/split.

212. *United States v. Cincinnati,* Memorandum of Agreement, (April 12, 2002) www.usdoj.gov/crt/split.

213. Merrick Bobb, Special Counsel, *15th Semiannual Report* (Los Angeles: Los Angeles Sheriff's Department, 2002), 99.

214. *Cincinnati Enquirer,* "2003: The Year in Review," December 31, 2003.

215. Milton and others, *Police Use of Deadly Force,* 134.

216. United States Department of Justice and the Metropolitan Police Department of the District of Columbia, *Memorandum of Agreement* (June 13, 2001), Par. 37, 53. Available at www.usdoj.gov/crt/split.

217. *United States v. City of Los Angeles,* Consent Decree, (June 15, 2001), Sec. III(A). "The OHB Unit shall have the capability to 'roll out' to all Categorical Use of Force incidents 24 hours a day. The Department shall require immediate notification to the Chief of Police, the OHB Unit, the Commission and the Inspector General by the LAPD whenever there is a Categorical Use of Force. Upon receiving each such

notification, an OHB Unit investigator shall promptly respond to the scene of each Categorical Use of Force and commence his or her investigation. The senior OHB Unit manager present shall have overall command of the crime scene and investigation at the scene where multiple units are present to investigate a Categorical Use of Force incident; provided, however, that this shall not prevent the Chief of Police, the Chief of Staff, the Department Commander or the Chief's Duty Officer from assuming command from a junior OHB supervisor or manager when there is a specific need to do so."

218. *United States v. City of Cincinnati,* Memorandum of Agreement, Par. 56.

219. The operations of this program are described in Los Angeles Sheriff's Department, Office of Independent Review, *Second Annual Report 2003* (Los Angeles: Los Angeles Sheriff's Department, 2003), 1–3. Available at www.laoir.org.

220. State Attorney's Office [Miami-Dade County, Florida], *The Rap Sheet* (November 2001). Available at www.state.fl.us.

221. *United States v. City of Los Angeles,* Consent Decree, Par. 61.

222. Ibid., Par. 60.

223. *People of California v. City of Riverside,* Stipulated Judgment (March 2001), Par. 58. Available at www.ci.riverside.ca.us/rpd.

224. Ibid., Par. 62.

225. James G. Kolts, *The Los Angeles Sheriff's Department* (Los Angeles: Los Angeles Sheriff's Department, 1992), 100. Available at www.parc.info.

226. *NAACP v. City of Philadelphia,* Plaintiffs' First Monitoring Report: Complaints Against Police (1997), 15.

227. *United States v. City of Cincinnati,* Memorandum of Agreement, Sec. IV(B) Par. 26–29.

228. The first study to identify the code of silence was William A. Westley, *Violence and the Police* (Cambridge: MIT Press, 1970).

229. *United States v. City of Los Angeles,* Consent Decree.

230. Police Whistleblowers Conference, Rutgers Camden Law School, April 2, 2004.

231. Kevin Keenan and Samuel Walker, *The Law Enforcement Officer's Bill of Rights: An Impediment to Accountability?* (Unpublished manuscript, 2004).

232. Coleen Kadleck and Samuel Walker, *An Analysis of the Accountability Related Provisions of Police Collective Bargaining Agreements* (Unpublished manuscript, 2004).

233. Police Assessment Resource Center, *Portland Police Bureau: Officer-Involved Shootings and In-Custody Deaths,* 50.

234. Ibid., 45.

235. Ibid., 134, 137–8. The Department of Justice endorses this idea: U.S. Department of Justice, *Principles for Promoting Police Integrity,* 5.

236. Ibid.

237. Department of Justice, *Principles for Promoting Police Integrity,* 6.

238. *United States v. City of Los Angeles*, Consent Decree.

239. Samuel Walker, *The Discipline Matrix: An Effective Police Accountability Tool?* (Omaha: University of Nebraska at Omaha, 2004). Available at www.police accountability.org.

240. *NAACP v. Philadelphia, Stipulated Agreement* (1996).

241. Philadelphia, Integrity and Accountability Office, *Use of Force,* 10–11.

242. Ibid., 12.

243. Ibid., 47.

244. Ibid., 28.

245. Ibid., 33, 36.

246. Ibid., 57, 61.

247. Philadelphia, Integrity and Accountability Office, *Discipline System*, 6.

248. Ibid., 52.

249. Oakland Police Department, *Negotiated Settlement Agreement, Second Semiannual Report* (February 18, 2004), 7–8. Available at www.oaklandpolice.com.

250. United States Justice Department, *Principles for Promoting Police Integrity*, 7. www.ncjrs.org, NCJ 186189.

251. President's Commission on Law Enforcement and Administration of Justice, *The Challenge of Crime in a Free Society* (New York: Avon Books, 1967). National Advisory Commission on Civil Disorders, *Report* (New York: Bantam Books, 1968).

252. See, for example, ACLU of Northern California, *Failing the Test: Oakland's Police Complaint Process in Crisis* (San Francisco: ACLU of Northern California, 1996) and U.S. Commission on Civil Rights, Wisconsin Advisory Committee, *Police Protection of the African American Community in Milwaukee* (Washington: Government Printing Office, 1994), 43–46. For a comprehensive account, see *Samuel Walker, Police Accountability: The Role of Citizen Oversight* (Belmont: Wadsworth, 2001).

253. James G. Kolts, *The Los Angeles Sheriff's Department* (Los Angeles: Sheriff's Department, 1992), 100. Available at www.parc.info.

254. LAPD, *Report of the Independent Monitor, Report for the Quarter Ending June 30, 2003* (2003), 3. Available at www.lapdonline.org.

255. This issue is discussed in detail in Walker, *Police Accountability: The Role of Citizen Oversight.*

256. The Web site of the Boise Ombudsman is www.boiseombudsman.org. This author published a Model Citizen Complaint Procedure in Walker, *Police Accountability: The Role of Citizen Oversight*, 188–97.

257. International Association of Chiefs of Police, *Investigation of Employee Misconduct.* Concepts and Issues Paper, rev. ed. (Gaithersburg, MD: IACP, July 2001).

258. Walker, *Police Accountability: The Role of Citizen Oversight,* Appendix, 188–97.

259. The best policies and procedures, cited throughout this chapter, have been developed by the San Jose Independent Police Auditor (www.ci.san-jose.ca.us/ipa), the

Washington, DC Office of Citizen Complaint Review (www.occr.dc.gov), and the Boise Community Ombudsman (www.boiseombudsman.org).

260. Debra Livingston, "Police Reform and the Department of Justice: An Essay on Accountability," *Buffalo Criminal Law Review* 2 (1999): 843.

261. Web sites: Portland: www.portlandonline.com/police, Washington, DC: http://mpdc.dc.gov, Springfield, MO: www.ci.springfield.mo.us/police.

262. Samuel Walker and Eileen Luna, *An Evaluation of the Oversight Mechanisms of the Albuquerque Police Department* (Albuquerque: City Council, 1997).

263. *United States v. New Jersey,* Consent Decree (December 30, 1999), Par. 59. Available at www.usdoj.gov/crt/split.

264. New Jersey, *Monitors' Seventh Report* (January 17, 2003), 75. Available at www.state.nj.us/lps/.

265. *United States v. Cincinnati,* Memorandum of Agreement (April 12, 2002). Available at www.usdoj.gov.crt/split.

266. Washington, DC, Office of Civilian Complaint Review. Available at www.occr.dc.gov.

267. Hervey Juris and Peter Feuille, *Police Unionism* (Lexington, MA: Lexington Books, 1977).

268. *United States v. Cincinnati,* Memorandum of Agreement (April 12, 2002), Par. 36. Available at www.usdoj.gov/crt/split.

269. *United States v. New Jersey,* Consent Decree (1999), Par. 59. Available at www.usdoj.gov/crt/split.

270. *United States v. the City of Los Angeles,* Consent Decree (June 15, 2001), Par. 74. Available at www.usdoj.gov/crt/split.

271. Ibid., Par. 74(a).

272. *United States v. the City of Los Angeles,* Consent Decree, Par. 74(b).

273. Minneapolis, Civilian Review Authority, *Annual Report 1998* (Minneapolis: CRA, 1999), Exhibit A, 2.

274. Washington, DC, Office of Citizen Complaint Review, *Annual Report 2003* (2003), 16. Available at www.occr.dc.gov.

275. Boise Community Ombudsman, *Policies and Procedures* (January 1, 2001), 6. Available at www.boiseombudsman.org.

276. IACP, *Investigation of Employee Misconduct.*

277. The Police Complaints Center that conducted the NYPD survey no longer operates its Web site. The incident is described in Walker, *Police Accountability: The Role of Citizen Oversight.*

278. This observation is based on the author's conversations with staff at several oversight agencies, including the now-abolished Minneapolis Civilian Review Authority.

279. Boise Community Ombudsman, *Policies and Procedures,* 4. Available at www.boiseombudsman.org.

280. San Jose Independent Police Auditor, *Year End Report, 1993–1994* (San Jose: IPA, 1995). San Jose Independent Police Auditor, *1995 Year End Report* (San Jose: IPA, 1996). Available at www.ci.san-jose.ca.us/ipa.

281. New York City Civilian Complaint Review Board, *Status Report January–June 2003*, (2003), Table 24 A, 57. Available at www.ci.nyc.ny.us/ccrb/.

282. Ibid.

283. San Jose, Independent Police Auditor, *Policy and Procedures*. Available at www.ci.san-jose.ca.us/ipa.

284. The original academic study that identified the code of silence is William A. Westley, *Violence and the Police* (Cambridge: MIT Press, 1970), based on a 1950 study of the Gary, Indiana, Police Department. A recent and more thorough report is David Weisburd and others, *The Abuse of Authority: A National Study of Police Officers' Attitudes* (Washington, DC: The Police Foundation, 2001). An executive summary is available at www.ncjrs.org, NCJ 181312.

285. Boise Community Ombudsman, *Policies and Procedures*, 9, 2.08. "Truthfulness and Cooperation." Available at www.boiseombudsman.org

286. *United States v. the City of Los Angeles,* Consent Decree, Par. 61.

287. San Francisco Office of Citizen Complaints, *Response to the Board of Supervisors Regarding SFPD's Patterns of Withholding Information Requested for OCC Investigations* (April 23, 2003). Available at www.ci.sf.ca.us/occ/. The San Francisco City Charter mandates that the Police Department provide the OCC full and prompt cooperation. Sec. 4.127 states, "In carrying out its objectives, the Office of Citizen Complaints shall receive prompt and full cooperation and assistance from all departments, officers and employees of the City and County. The director may also request and the Chief of Police shall require the testimony or attendance of any member of the Police Department."

288. San Francisco Office of Citizen Complaints, *Response to the Board of Supervisors Regarding SFPD's Patterns of Withholding Information Requested for OCC Investigations.*

289. *United States v. New Jersey,* Consent Decree, Par. 77.

290. San Jose, Independent Police Auditor, *Mid-Year Report, June 1995* (1995), 19–20. Available at www.ci.san-jose.ca.us/ipa.

291. Omaha Public Safety Auditor, *Public Safety Auditor's Report for the Quarter Ending June 30th 2003* (2003), 14. Available at www.ci.omaha.ne.us.

292. Pittsburgh Police Bureau , *Auditor's Eighteenth Quarterly Report. Quarter Ending February 16, 2002* (2002), p. 53.

293. Richard Jerome, *Police Oversight Project—City of Albuquerque* (Los Angeles: Police Assessment Resource Center, 2002), 34–35. Available at www.parc .info.

294. Author's interviews with Citizens Police Review Board staff, 2001.

295. San Jose, Independent Police Auditor, *Policy and Procedures*. Available at www.ci.san-jose.ca.us/ipa.

296. *United States v. New Jersey,* Consent Decree, Par. 75.

297. San Jose, Independent Police Auditor, *Policy and Procedures*.

298. Los Angeles Sheriff's Department, Office of Independent Review, *Second Annual Report 2003* (Los Angeles: Sheriff's Department, 2003), 77–78. Available at www.laoir.com.

299. Amy Oppenheimer and Craig Pratt, *Investigating Workplace Harassment: How to be Fair, Thorough, and Legal* (Alexandria, VA: Society for Human Resource Management, 2003), 110–11.

300. *United States v. Cincinnati,* Memorandum of Agreement, Par. 41.

301. *United States v. the City of Los Angeles,* Consent Decree, Par. 84. Available at www.lapdonline.org.

302. *People of California v. City of Riverside,* Stipulated Judgment (March 5, 2001), Par. 50. Available at www.ci.riverside.ca.us/rpd.

303. San Francisco Office of Citizen Complaints, *Year 2001 Annual Report, 6,* available online at www.ci.sf.ca.us/occ/. "During 2001, OCC identified an average of 4.67 allegations per civilian complaint (4250 allegations in 911 complaints filed, excluding merged, voided and no finding cases). In 2001, as in the four previous years, by the measure of average number of allegations identified, OCC maintained the previously documented level of improvement in completeness of its analysis of complaints."

304. New York City Civilian Complaint Review Board, *Status Report, January–June 2003, 9.*

305. *United States v. the City of Los Angeles,* Consent Decree, Par. 82.

306. San Jose, Independent Police Auditor, *Policy and Procedure.* Available at www.ci.san-jose.ca.us/ipa.

307. Boise Community Ombudsman, *Policies and Procedures, 9.* "2.09 Tape Recordings. a. The complete interview of an officer/employee accused of a Class I violation shall be recorded and a copy may be obtained by the officer/employee under investigation upon request. The officer/employee may also bring his/her own recording device, if he/she wishes. The cost of taping and any mechanical devices used by the officer/ employee shall be borne by the officer/employee." www.boiseombudsman.org.

308. *United States v. the City of Los Angeles,* Consent Decree, Par. 80.

309. The former Police Internal Investigations Advisory Committee (PIIAC) has since been replaced by the Independent Police Review Division, housed in the Office of the City Auditor. Available at www.ci.portland.or.us/auditor.

310. San Jose Independent Police Auditor, *Mid-Year 2003 Report, 8.*

311. "A Model Citizen Complaint Procedure," in Walker, *Police Accountability: The Role of Citizen Oversight,* Appendix, 188–97.

312. San Jose Independent Police Auditor, *Year-End Report 1994, 16.*

313. Amy Oppenheimer and Craig Pratt, *Investigating Workplace Harassment: How to be Fair, Thorough, and Legal* (Alexandria, VA: Society for Human Resource Management, 2003), 108.

314. *United States v. New Jersey,* Consent Decree, Par. 81. "The State shall make findings based on a "preponderance of the evidence" standard."

315. Pittsburgh, *Auditor's Eighteenth Quarterly Report. Quarter Ending February 16, 2002, 26.*

316. Philadelphia, Integrity and Accountability Office, *Disciplinary System* (Philadelphia: Philadelphia Police Department, 2001), 53.

317. Washington, DC, Office of Citizen Complaint Review, Press Release, August 5, 2003. Available at www.occr.dc.org.

318. New York Civil Liberties Union, *Five Years of Civilian Review: A Mandate Unfulfilled* (New York: NYCLU, 1998).

319. Portland, Police Internal Investigations Auditing Committee, *Fourth Quarter Monitoring Report 1997* (Portland, PIIAC, 1998), 11.

320. Commission on Accreditation for Law Enforcement Agencies, *Standards for Law Enforcement Agencies,* 4th ed. (Fairfax, VA: CALEA, 1998).

321. San Francisco, Office of Citizen Complaints, *2001 Annual Report.*

322. Ibid.

323. Oakland Police Department, *Negotiated Settlement Agreement Second Semiannual Report* (February 2004), Task 02, Sec. III.B.2, p. 22. Available at www .oaklandpolice.com.

324. San Jose Independent Police Auditor, *1995 Year End Report,* 21–22.

325. These documents are readily accessible on the Web: www.ci.san-jose .ca.us/ipa/home.html. www.boiseombudsman.org. www.occr.dc.gov.

326. U.S. Department of Justice, *Investigation of the Portland, Maine, Police Department. Letter to Mr. Gary Wood, Corporation Counsel,* March 21, 2003.

327. Jayson Wechter, *Investigating Police Misconduct is Different* (2004). Available at www.policeaccountability.org.

328. Samuel Walker, Carol Archbold, and Leigh Herbst, *Mediating Citizen Complaints Against Police Officers: A Guide for Police and Community Leaders* (Washington, DC; Department of Justice, 2002). Available at www.cops.usdoj.gov.

329. Oakland Police Department, *Negotiated Settlement Agreement Second Semi-annual Report,* vii.

330. ACLU—National Capital Area, *Analysis of the District of Columbia's Civilian Complaint Review Board and Recommendations for its Replacement* (Washington: ACLU-National Capital Area, 1995).

331. San Jose, Independent Police Auditor, *Policy and Procedures.* Available at www.san-jose.ca.us/ipa.

332. New Jersey, *Monitors' Seventh Report* (January 17, 2003), 75. Available at www.state.nj.us/lps/

333. Minneapolis Civilian Review Authority, *1999 Annual Report* (Minneapolis: CRA, 1999), Exhibit C.

334. Michele Sviridoff and Jerome E. McElroy, *Processing Complaints Against Police in New York City: The Complainant's Perspective* (New York: Vera Institute of Justice, January 1989). *The Processing of Complaints Against Police in New York City: The Perceptions and Attitudes of Line Officers* (New York: Vera Institute of Justice, September 1989).

335. Pittsburgh, *Auditor's Eighteenth Quarterly Report. Quarter Ending February 16, 2002,* 53.

336. Los Angeles Police Department, *Report of the Independent Monitor for the Los Angeles Police Department, Report for the Quarter Ending June 30, 2003* (Los Angeles: Los Angeles Police Department, 2003), 3–4.

337. *People v. City of Riverside,* Stipulated Judgment, Par. 51. Available at www .ci.riverside.ca.us.

338. ACLU of Northern California, *Failing the Test: Oakland's Police Complaint Process in Crisis* (San Francisco: ACLU of Northern California, 1996).

339. The issues surrounding the sustain rate are discussed at length in Walker, *Police Accountability: The Role of Citizen Oversight,* 120–22, 134–35.

340. U.S. Commission on Civil Rights, *Who is Guarding the Guardians? A Report on Police Practices* (Washington, DC: Commission on Civil Rights, 1981), 166.

341. Christopher Commission, *Report of the Independent Commission on the Los Angeles Police Department* (Los Angeles: Christopher Commission, 1991). Available at www.parc.info.

342. James G. Kolts, *The Los Angeles Sheriff's Department* (Los Angeles: Los Angeles Sheriff's Department, 1992). Available at www.parc.info.

343. U.S. Commission on Civil Rights, *Who is Guarding the Guardians? A Report on Police Practices,* 159, 166.

344. Samuel Walker, *Early Intervention Systems for Law Enforcement Agencies: A Planning and Management Guide* (Washington, DC: Department of Justice, 2003). Available at www.ncjrs.org, NCJ 201245.

345. Robert C. Davis, Christopher Ortiz, Nicole J. Henderson, Joel Miller, and Michelle K. Massie, *Turning Necessity into Virtue: Pittsburgh's Experience with a Federal Consent Decree* (New York: Vera Institute of Justice, 2002), 37. Available at www.vera.org, or www.ncjrs.org, NCJ 200251.

346. Samuel Walker, Geoffrey P. Alpert, and Dennis Kenney. *Early Warning Systems: Responding to the Problem Police Officer,* Research in Brief (Washington, DC: National Institute of Justice, 2001). Available at www.ncjrs.org, NCJ 188565.

347. International Association of Chiefs of Police, *Building Integrity and Reducing Drug Corruption in Police Departments* (Washington, DC: Government Printing Office, 1989), 80.

348. Comments, Commander responsible for EI system in an unidentified police department. Early Intervention Systems, State of the Art Conference, Phoenix, AZ, January 2003.

349. The issue of terminology is discussed in Walker, *Early Intervention Systems for Law Enforcement Agencies,* 8–9.

350. Herman Goldstein, *Police Corruption* (Washington: The Police Foundation, 1975).

351. It is significant, for example, that a recent National Institute of Justice (NIJ) publication on developing programs to deal with law enforcement officer stress includes a section on "Selecting Target Groups" but contains no reference to specific performance indicators such as are commonly used in EI systems. Peter Finn and Julie Esselman Tomz, *Developing a Law Enforcement Stress Program for Officers and Their Families* (Washington: Government Printing Office, 1997), 23–26. Available at www.ncjrs.org, NCJ163175.

352. Frank Landy, *Performance Appraisal in Police Departments* (Washington, DC: The Police Foundation, 1977).

353. Gary Stix, "Bad Apple Picker: Can a Neural Network Help Find Problem Cops?" *Scientific American* (December 1994): 44–45. Bernard Cohen and Jan M. Chaiken,

Police Background Characteristics and Performance (Lexington, MA: Lexington Books, 1973).

354. A good description of these two systems is available in Special Counsel Merrick J. Bobb, *16th Semiannual Report* (Los Angeles: Special Counsel, 2003), 68–72. Available at www.parc.info.

355. Comments, unidentified police chiefs, Justice Department Conference, *Strengthening Police Community Relationships*, June 9–10, 1999.

356. Herman Goldstein, *Policing a Free Society* (Cambridge, MA: Ballinger, 1977), 171. He cited an experimental (but short-lived) program by Hans Toch in the 1970s in which peer officers counseled Oakland, California, police officers with records of use of force incidents.

357. Catherine H. Milton, Jeanne Wahl Halleck, James Lardner, and Gary L. Abrecht, *Police Use of Deadly Force* (Washington, DC: The Police Foundation, 1977).

358. U.S. Commission on Civil Rights, *Who is Guarding the Guardians?* 81–86.

359. New York City, Commission to Investigate Allegations of Police Corruption and the Anti-Corruption Procedures of the Police Department, [Mollen Commission], *Commission Report* (New York, 1994). Available at www.parc.info.

360. *Allen v. City of Oakland* (2003). The consent decree is available at www.oaklandpolice.com.

361. Bruce Porter, *The Miami Riot of 1980* (Lexington, MA: Lexington Books, 1984). U.S. Commission on Civil Rights, *Confronting Racial Isolation in Miami* (Washington, DC: Government Printing Office, 1982).

362. Christopher Commission, *Report of the Independent Commission on the Los Angeles Police Department* (Los Angeles: The Christopher Commission, 1991). Available at www.parc.info.

363. Ibid., 40–48.

364. Kolts, *The Los Angeles County Sheriff's Department.* Available at www.parc.info.

365. U.S. Department of Justice, *Principles for Promoting Police Integrity* (Washington, DC: U.S. Department of Justice, 2001). Available at www.ncjrs.org, NCJ 186189.

366. The various consent decrees and memoranda of understanding, along with other related documents, are available at www.usdoj.gov/crt/split.

367. Commission on Accreditation for Law Enforcement Agencies, *Standards for Law Enforcement Agencies,* 4th ed. , Standard 35.1.15, "Personnel Early Warning System" (2001).

368. Samuel Walker, Geoffrey P. Alpert, and Dennis J. Kenney, *Early Warning Systems: Responding to the Problem Police Officer* (Washington, DC: Government Printing Office, 2001), described a three-component approach. The four-component approach used here reflects the subsequent research. Available at www.ncjrs.org, NCJ 188565.

369. Walker, Alpert, and Kenney, *Early Warning Systems: Responding to the Problem Police Officer.*

370. Samuel Walker, *Police Accountability: The Role of Citizen Oversight* (Belmont: Wadsworth, 2001), 121–35.

371. *Allen v. City of Oakland* (2003). The consent decree is available at www.oaklandpolice.com.

372. Paul Chevigny, "Force, Arrest, and Cover Charges," in *Police Power: Police Abuses in New York City* (New York: Vintage Books, 1969), 136–46.

373. Bobb, *16th Semiannual Report.*

374. Walker, Alpert, and Kenney, *Early Warning Systems: Responding to the Problem Police Officer.*

375. Special Counsel Merrick J. Bobb, *15th Semiannual Report* (Los Angeles: Los Angeles Sheriff's Department, 2002), 39, 42. Available at www.parc.info.

376. This approach is described in Special Counsel Merrick Bobb, *16th Semiannual Report,* (Los Angeles: Los Angeles Sheriff's Department, 2003), 74–75. Available at www.parc.info.

377. Davis, Ortiz, Henderson, Miller, and Massie, *Turning Necessity into Virtue: Pittsburgh's Experience with a Federal Consent Decree,* 45–47. Available at www .ncjrs.org, NCJ 200251. (see n. 345)

378. *United States v. Cincinnati,* Memorandum of Agreement, Sec. IA (A), www.usdoj.gov/crt/split.

379. Lorie Fridell, *By the Numbers: Analyzing Race Data* (Washington, DC: Police Executive Research Forum, 2004).

380. *United States v. Pittsburgh,* Consent Decree, Paragraph 20-b, www.usdoj .gov/crt/split.

381. Pittsburgh Police Bureau, *Auditor's Eighteenth Quarterly Report, Quarter Ending February 16, 2002,* 18.

382. Omaha Police Union, "Bad Boy/Girl Class Notes Shared," *The Shield* (April 1992). A commander with the Kansas City Police Department told the author about a counterproductive experience in that department. Interview, March 2004.

383. Miami-Dade Police Department, Employee Identification System. Excerpts in Walker, *Early Identification Systems for Law Enforcement Agencies,* Appendix C.

384. This topic was the subject of a working conference cosponsored by the Austin, TX, Police Department and this author, March 3, 2004.

385. Discussion, Early Intervention System State of the Art Conference, Phoenix, AZ, February,2003.

386. Walker, Alpert, and Kenney, *Early Warning Systems: Responding to the Problem Police Officer.*

387. Davis and others, *Turning Necessity into Virtue.*

388. Walker, Alpert, and Kenney, *Early Warning Systems: Responding to the Problem Police Officer.*

389. Pittsburgh Police Bureau, *Monitor's Eighteenth Quarterly Report, Quarter Ending February 16, 2002,* 16.

390. Pittsburgh Police Bureau, *Monitor's Eighteenth Quarterly Report, Quarter Ending February 16, 2002,* 21.

391. These quotes and a full discussion of the survey are from Walker, *Early Intervention Systems for Law Enforcement Agencies,* chap. 4.

392. Los Angeles Police Department, *10th Quarterly Report of the Independent Monitor,* 10. Par. 51c. Available at www.lapdonline.org.

393. Davis and others, *Turning Necessity into Virtue.*

394. *United States v. City of Los Angeles,* Consent Decree (2000), Sec. II, Par. 47.

395. Robin Sheppard Engel, *How Police Supervisory Styles Influence Patrol Officer Behavior* (Washington, DC: U.S. Department of Justice, 2003). Available at www.ncjrs.org, NCJ 194078.

396. Early Intervention System State of the Art Conference, *Report* (Omaha, 2003). The essence of the report is in Walker, *Early Intervention Systems for Law Enforcement Agencies.*

397. San Jose, Independent Police Auditor, *2001 Year End Report,* 48–49. Available at www.san-jose.ca.us/ipa.

398. Herman Goldstein, "Improving Policing: A Problem-Oriented Approach," *Crime and Delinquency* 25 (1979): 236–58. Michael S. Scott, *Problem-Oriented Policing: Reflections on the First 20 Years* (Washington, DC: U.S. Department of Justice, 2000). Available at www.ncjrs.org.

399. James J. Willis, Stephen D. Mastrofski, David Weisburd, and Rosann Greenspan, *Compstat and Organizational Change in the Lowell Police Department* (Washington, DC: The Police Foundation, 2004).

400. Davis and others, *Turning Necessity into Virtue: Pittsburgh's Experience with a Federal Consent Decree* (New York: Vera Institute of Justice, 2002), 45–47. Available at www.ncjrs.org, NCJ 200251.

401. Carol Archbold, *Police Accountability, Risk Management, and Legal Advising* (New York: LFB Scholarly Publishing, 2004). Michel Crouhy, Dan Galai, and Robert Mark, *Risk Management* (New York: McGraw-Hill, 2001).

402. Walker, Alpert, and Kenney, *Early Warning Systems: Responding to the Problem Police Officer.*

403. Davis and others, *Turning Necessity into Virtue.*

404. Samuel Walker, Geoffrey P. Alpert, and Dennis J. Kenney, *Responding to the Problem Police Officer,* Final Report to the National Institute of Justice.

405. Special Counsel Merrick J. Bobb, *15th Semiannual Report* (Los Angeles: Special Counsel, 2002), 68. Available at www.parc.info.

406. Ibid., 64.

407. The survey is reported in detail in Walker, *Early Intervention Systems for Law Enforcement Agencies,* chap. 4.

408. See the discussion of this issue in the evaluation of the Pittsburgh consent decree. Davis and others, *Turning Necessity into Virtue.*

409. In Seattle the union demanded that the EI system be subject to collective bargaining or meet and confer. The issue remains unresolved at the time this report was written. Nonetheless, this remains the only known case of union opposition to the creation of an EI system.

410. The demand to make an EI system a subject for collective bargaining has occurred in Seattle, Washington. Most experts believe that an EI system is an administrative tool that is clearly a prerogative of management that should not be subject to collective bargaining.

411. Pennsylvania State Police, *Annual Report* (2002), 19. Available at www.psp.state.pa.us.

412. U.S. Department of Justice, *Investigation of the Miami Police Department, Letter to Alejandro Vilarello,* March 19, 2003, 19. Available at www.usdoj.gov.crt.split.

413. Richard Jerome, *Police Oversight Project. City of Albuquerque* (Los Angeles: Police Assessment Resource Center, 2002), 80–81. Available at www.parc.info.

414. The chronology of events is very well documented in Los Angeles Police Commission, *Report of the Rampart Independent Review Panel* (Los Angeles: Los Angeles Police Commission, 2000), 136–62 and Appendix B, 217–19. Available at www.lapdonline.org.

415. Christopher Commission, *Report of the Independent Commission on the Los Angeles Police Department* (Los Angeles: City of Los Angeles, 1991). Available at www.parc.info.

416. Merrick Bobb, *Five Years Later: A Report to the Los Angeles Police Commission* (Los Angeles: Los Angeles Police Commission, 1996). Available at www.parc.info.

417. *United States v. the City of Los Angeles,* Consent Decree (2000). The consent decree is available at www.usdoj.gov/crt/split.

418. Los Angeles Police Department, *10th Quarterly Report of the Independent Monitor,* 6. Available at www.lapdonline.org.

419. Special Counsel Merrick J. Bobb, *11th Semiannual Report* (Los Angeles: Los Angeles Sheriff's Department, 1999), 55. Available at www.parc.info.

420. Bobb, *16th Semiannual Report,* 44, 49.

421. Ibid., 57–58.

422. Ibid., 43, 58.

423. Pittsburgh Police Bureau, *Monitor's Eighteenth Quarterly Report, Quarter Ending February 16, 2002,* 6.

424. Davis and others, *Turning Necessity into Virtue.*

425. Samuel Walker, *Police Accountability: The Role of Citizen Oversight.*

426. The evidence on the effectiveness of citizen oversight agencies is somewhere between weak and nonexistent. For a discussion of the related issues, see Walker, *Police Accountability: The Role of Citizen Oversight.*

427. Los Angeles Sheriff's Department, Office of Independent Review, *First Annual Report 2002* (Los Angeles: Los Angeles Sheriff's Department, 2002), 32. Available at www.laoir.com.

428. Samuel Walker, "Setting the Standards: The Efforts and Impacts of Blue-Ribbon Commissions on the Police," in *Police Leadership in America: Crisis and Opportunity,* ed. William A. Geller, 354–70 (New York: Praeger, 1985).

429. The nature of local political cultures and how they support or discourage police accountability is an extremely important subject that has not been adequately studied.

430. On the fight over citizen oversight in San Jose, see ACLU of Northern California, *A Campaign of Deception: San Jose's Case Against Civilian Review* (San Francisco: ACLU of Northern California, 1992). Walker, *Police Accountability: The Role of Citizen Oversight*, 38–40.

431. The ordinances establishing police auditors are generally available on their Web sites. These Web sites are conveniently available at www.policeaccountability.org.

432. The enabling ordinance is available on the IPA's Web site: http://www.ci .san-jose.ca.us/ipa/home.html.

433. Please see www.boiseombudsman.org and www.ci.omaha.ne.us.

434. Please see www.parc.info and www.laoir.com.

435. *NAACP, ACLU, the Barrio Project v. City of Philadelphia, Settlement Agreement* (1996).

436. Seattle Citizens Review Panel, *Final Report* (Seattle: Office of the Mayor, 1999).

437. San Jose Independent Police Auditor, *Year End Report, 1993–1994* (San Jose: Independent Police Auditor, 1994). Available at www.ci.san-jose.us.ca/ipa/home.html.

438. Philadelphia Office of Integrity and Accountability, *Enforcement of Narcotics Laws* (Philadelphia: Philadelphia Police Department, July 2002), 1–2.

439. Walker, *Police Accountability: The Role of Citizen Oversight*, 93–104.

440. San Jose Independent Police Auditor, *Year End Report 2002*, Appendix F, 83–92. Available at www.ci.san-jose.ca.us/ipa.

441. Merrick Bobb, Special Counsel, *15th Semiannual Report* (Los Angeles: Los Angeles Sheriff's Department, 2002), 14. Available at www.parc.info.

442. San Jose, Independent Police Auditor, *A Student's Guide to Police Practices* (San Jose: Independent Police Auditor, 2002). Available at www.ci.san-jose.ca.us/ipa.

443. Seattle Police Department, Office of Professional Accountability, *Respect: Voices and Choices*. Available at www.cityofseattle.net/police.

444. Please see www.ci.san-jose.ca.us/ipa.

445. Please see www.boiseombudsman.org.

446. Archived at www.parc.info.

447. Please see www.ci.nyc.ny.us/html/ccrb/home.html.

448. Please see www.kcpd.org and www.phila.gov/pac.

449. See, for example, Los Angeles Sheriff's Department, Office of Independent Review, *Report of Oversight of Administrative Discipline Cases: October thru December 2003* (Los Angeles: LASD, 2004). Available at www.laoir.com.

450. Samuel Walker, "Setting the Standards: The Efforts and Impacts of Blue-Ribbon Commissions on the Police," in *Police Leadership in America: Crisis and Opportunity*, ed. W. A. Geller, 354–70 (New York: Praeger, 1985). See also the valuable collection of reports in *The Politics of Riot Commissions*, ed. Anthony M. Platt (New York: Collier Books, 1971).

451. A very useful collection of excerpts from riot commission reports, together with commentary, is Platt, *The Politics of Riot Commissions.*

452. The reports of monitors are generally available on the Web sites of the police departments in question.

453. Such boards are classified as Class II forms of oversight in Walker, *Police Accountability: The Role of Citizen Oversight,* 62–63.

454. These issues are discussed in greater detail in Walker, *Police Accountability: The Role of Citizen Oversight.*

455. Anthony M. Pate and Lorie A. Fridell, *Police Use of Force,* 2 Vols. (Washington, DC: The Police Foundation, 1993).

456. Walker, *Police Accountability,* chap. 5.

457. Barbara Armacost, "Organizational Culture and Police Misconduct," *George Washington Law Review* 72 (March 2004): 493.

458. The best literature on police internal affairs units at this point is found not in the academic literature but in what can be called the professional literature, primarily the reports of the police auditors described in this chapter and in the reports of police monitors.

459. This point is argued in Walker, *Police Accountability: The Role of Citizen Oversight,* with more extensive material on other citizen oversight agencies. The reports of the Special Counsel are archived at www.parc.info.

460. All the reports are available at www.parc.info.

461. Available at www.parc.info.

462. James G. Kolts, *The Los Angeles Sheriff's Department* (Los Angeles: Los Angeles Sheriff's Department, 1992). Available at www.parc.info.

463. Kolts, *The Los Angeles Sheriff's Department,* 75.

464. Merrick Bobb, Special Counsel, *1st Semiannual Report* (Los Angeles: Los Angeles Sheriff's Department, 1993), 69–75. Available at www.parc.info.

465. Merrick Bobb, Special Counsel, *11th Semiannual Report* (Los Angeles: Los Angeles Sheriff's Department, 1999). Available at www.parc.info.

466. *United States v. Cincinnati,* Memorandum of Agreement, 2002, Sec. IV (C). Available at www.usdoj.gov/crt/split.

467. Merrick Bobb, Special Counsel, *14th Semiannual Report* (Los Angeles: Los Angeles Sheriff's Department, 2001), 93–104. Kolts, *The Los Angeles Sheriff's Department,* 25–26.

468. Merrick Bobb, Special Counsel *16th Semiannual Report* (Los Angeles: Los Angeles Sheriff's Department, 2003), 109. Available at www.parc.info.

469. Merrick Bobb, Special Counsel, *9th Semiannual Report* (Los Angeles: Los Angeles Sheriff's Department, 1998), 12. Available at www.parc.info.

470. Bobb, *9th Semiannual Report,* 8.

471. "Moreover, the sergeant to-patrol deputy ratio at Century ranges over time from about 20–25 to 1, whereas the station management believes that an 8-1 ratio would be optimal." Bobb, *9th Semiannual Report,* 23.

472. Bobb, *15th Semiannual Report,* 26.

473. Ibid., 9.

474. Ibid., 9.

475. Bobb, *16th Semiannual Report.*

476. Merrick Bobb, Special Counsel, *2nd Semiannual Report* (Los Angeles: Los Angeles Sheriff's Department, 1994), 34–35.

477. Bobb, *16th Semiannual Report*, 5–41.

478. See, for example, Los Angeles Sheriff's Department, Office of Independent Review, *Report of Oversight of Administrative Discipline Cases: October thru December 2003* (2004). Available at www.laoir.com.

479. Los Angeles Sheriff's Department, Office of Independent Review, *Second Annual Report 2003* (Los Angeles: Los Angeles Sheriff's Department, 2003), 71–73.

480. Ibid., iii.

481. Ibid., 2.

482. Ibid., 62–63.

483. Ibid., 63

484. Philadelphia Office of Integrity and Accountability, *Disciplinary System* (Philadelphia: Philadelphia Police Department, 2001).

485. San Jose, Independent Police Auditor, *2001 Year End Report* (San Jose: Independent Police Auditor 2000), xvii. Available at www.ci.san-jose.ca.us/ipa.

486. San Jose Independent Police Auditor, *2000 Year End Report,* 42.

487. San Jose Independent Police Auditor, *2003 Year End Report,* 55–60.

488. San Jose Independent Police Auditor, *2001 Year End Report,* 20–30.

489. San Jose Independent Police Auditor, *A Student's Guide to Police Practices* (San Jose: Independent Police Auditor, 2002). Available at www.ci.san-jose.ca.us.

490. Boise Community Ombudsman, *Public Report: Police Handling of a Reported Rape in Barber Park* (Boise: Community Ombudsman, June 27, 2000). Available at www.boiseombudsman.org.

491. Seattle Police Department, Office of Professional Accountability, *Report on Seattle's Response to Concerns about Racially Biased Policing* (Seattle: Seattle Police Department, June 2003). Available at www.ci.seattle.wa.us/police/OPA/.

492. Ibid., 8.

493. Philadelphia Police Department, Integrity and Accountability Office, *Use of Force* (Philadelphia: Philadelphia Police Department, 1999). Philadelphia Police Department, Integrity and Accountability Office, *Disciplinary System* (Philadelphia: Philadelphia Police Department, 2001). Philadelphia Police Department, Integrity and Accountability Office, *Enforcement of Narcotics Law* (Philadelphia: Philadelphia Police Department, 2002).

494. Bobb, *15th Semiannual Report,* 9.

495. Knapp Commission, *Report* (New York: Brazillier, 1971). Lawrence W. Sherman, *Scandal and Reform* (Berkeley: University of California Press, 1978). Mollen Commission, *Report* (New York City: Mollen Commission, 1994). Available at www.parc.info.

496. Mollen Commission, *Report.* Available at www.parc.info.

497. Los Angeles Police Department, *Board of Inquiry Report on the Rampart Incident* (Los Angeles: Los Angeles Police Department, 1999). Available at www .lapdonline.org. Los Angeles Police Commission, *Report of the Independent Review Panel* (Los Angeles: Los Angeles Police Commission, 2000). Available at www.lapdon line.org.

498. John Crew, personal conversations with author. ACLU of Northern California, *A Campaign of Deception: San Jose's Case Against Civilian Review.*

499. The activities of Portland Copwatch are available at www.portlandcop watch.org.

500. This author's personal experience suggests that the Seattle auditor was willfully inactive. In the mid-1990s the Seattle auditor did not even respond to telephone requests for an interview or even copies of the quarterly reports. (Other citizen oversight officials across the country have always been more than eager to talk about their activities.)

501. Seattle, *Citizens Review Panel: Final Report* (August 19, 1999). Seattle Police Department, *Expectations, Initiative & Accomplishment: A Five Year Summary* (Seattle: Seattle Police Department, August 1999).

502. Samuel Walker and Eileen Luna, *A Report on the Oversight Mechanisms of the Albuquerque Police Department* (Albuquerque: City Council, 1997), "Executive Summary."

503. Samuel Walker and Eileen Luna, "Institutional Structure vs. Political Will: Albuquerque as a Case Study in the Effectiveness of Citizen Oversight of the Police," in *Civilian Oversight of Policing: Governance, Democracy and Human Rights,* ed. Andrew Goldsmith and Colleen Lewis, 83–104 (Oxford: Hart Publishing, 2000).

504. Los Angeles Sheriff's Department, Office of Independent Review, *First Annual Report 2002,* 1. Available at www.laoir.org.

505. Ibid., 2.

506. Robin Sheppard Engel, *How Police Supervisory Styles Influence Patrol Officer Behavior* (Washington, DC: U.S. Department of Justice, 2003). Available at www.ncjrs.org, NCJ 194078

507. Samuel Walker, *Early Intervention Systems for Law Enforcement Agencies: A Planning and Management Guide* (Washington, DC: Department of Justice, 2003). Available at www.ncjrs.org, NCJ 201245.

508. Barbara Armacost, "Organizational Culture and Police Misconduct," *George Washington Law Review* 72 (March 2004): 515.

509. National Academy of Sciences, *Fairness and Effectiveness in Policing: The Evidence* (Washington, DC: National Academy Press, 2004).

510. An extremely useful guide, which appeared just as this book was being completed is Maya Harris West, *Organized for Change: The Activist's Guide to Police Reform* (Oakland: Policylink, 2004). Available at www.policylink.org.

511. Philadelphia Police Department, Integrity and Accountability Office, *Disciplinary System* (Philadelphia: Philadelphia Police Department, 2001).

512. Merrick Bobb, Special Counsel, *16th Semiannual Report* (Los Angeles: Los Angeles Sheriff's Department, 2003), 43–60. Available at www.parc.info.

513. Los Angeles Police Department, *Eleventh Quarterly Report of the Independent Monitor* (2004), 5–6. Available at www.lapdonline.org.

514. Merrick Bobb, Special Counsel, *9th Semiannual Report* (Los Angeles: Los Angeles Sheriff's Department, 1998). Available at www.parc.info. *People v. City of Riverside,* Stipulated Judgment (2001), Sec. 52. Available at www.ci.riverside,ca.us/rpd.

515. Los Angeles Police Department, *Board of Inquiry Report on the Rampart Incident* (Los Angeles: Los Angeles Police Department, 1999). Available at www.lapdonline.org. Philadelphia Police Department, Integrity and Accountability Office, *Disciplinary System* (Philadelphia: Philadelphia Police Department, 2001), 54.

516. Omaha Public Safety Auditor, *Public Safety Auditor's Report* (Quarter Ending September 30, 2002), 16. Available at www.ci.omaha.ne.us.

517. New Jersey State Police, *Monitor's Ninth Report* (January 2004), 39. Available at www.njpublicsafety.com.

518. Oakland Police Department, *Negotiated Settlement Agreement, Second Semiannual Report* (February 2004), 10. Available at www.oaklandpolice.com.

519. Philadelphia Police Department, Office of Integrity and Accountability, *Disciplinary System* (2001).

520. Personal conversations, author and Omaha Public Safety Auditor. See the auditor's reports at www.ci.omaha.ne.us.

521. Omaha Public Safety Auditor, *Public Safety Auditor's Report, Quarter Ending September 30, 2002, 16.* Available at www.ci.omaha.ne.us.

522. Merrick Bobb, Special Counsel, *15th Semiannual Report* (Los Angeles: Los Angeles Sheriff's Department, 2002), 99. www.parc.info.

523. Philadelphia Police Department, Integrity and Accountability Office, *Disciplinary System,* 49–50. A span of control of 7:1 is mandated in the Riverside, California consent decree. *People v. City of Riverside,* Stipulated Judgment (2001), Sec. 52. Available at www.ci.riverside,ca.us/rpd.

524. Merrick Bobb, Special Counsel, *16th Semiannual Report* (Los Angeles: Los Angeles Sheriff's Department, 2003). Available at www.parc.info.

525. Malcolm M. Feeley and Edward L. Rubin, *Judicial Policy Making and the Modern State* (New York: Cambridge University Press, 1998).

526. Metropolitan Police Department for Washington, DC, *Fifth Quarterly Report of the Independent Monitor* (July 2003), 15. Available at www.policemonitor.org.

527. Ibid., 2.

528. Oakland Police Department, *Second Semi-Annual Report of the Monitor* (2004), 4. Available at www.oaklandpolice.com.

529. Ibid., 52.

530. Samuel Walker, *The Discipline Matrix: An Effective Police Accountability Tool?* (Omaha: University of Nebraska at Omaha, 2003). Available at www.police account ability.org.

531. *Washington Post,* "Exploding Number of SWAT Teams Sets Off Alarms," June 17, 1997. Dave Kopel, "Smash-up Policing: When law enforcement goes Military," *National Review,* May 22, 2000.

532. Fresno Police Department, *Study of Use of Force* (2002). Available at www.fresno.gov/fpd.

533. Fresno Police Department, *Reportable Use of Force Project, Fourth Quarter 2003* (2003). Available at www.fresno.gov/fpd.

534. Seattle Police Department, *Special Report: Use of Force by SPD Officers* (November 2001). Available at www.cityofseattle.gov.

535. Seattle Police Department, *SPD Progress Report—Less Lethal Options Program—Year 1* (May 2002). Available at www.cityofseattle.gov/police.

536. Police Executive Research Forum, *Racially Biased Policing: A Principled Response* (Washington, DC: PERF, 2001).

537. Seattle Police Department, Office of Professional Accountability, *Report on Seattle's Response to Concerns about Racially Biased Policing* (June 2003). Available at www.cityofseattle.gov/police.

538. Seattle Police Department, Office of Professional Accountability, *OPA's Internal Outreach Efforts* (2004). Available at www.cityofseattle.gov/police.

INDEX

ABOUT THE AUTHOR

Samuel Walker is Isaacson Professor of Criminal Justice at the University of Nebraska at Omaha. His research interests involve police accountability, including citizen oversight of the police, early intervention systems for police officers, and the mediation of citizen complaints against police officers.

Professor Walker is the author of 12 books on policing, criminal justice policy, and civil liberties. His most recent book is *The Police in America: An Introduction* (5th ed., 2005). He is also the author of *Police Accountability: The Role of Citizen Oversight* (2001), *Taming the System: The Control of Discretion in Criminal Justice, 1950–1990* (1993), *Sense and Nonsense About Crime* (5th ed., 2001), *The Color of Justice: Race, Ethnicity, and Crime in America* (with C. Spohn & M. DeLone) (3rd ed., 2003), and *In Defense of American Liberties: A History of the ACLU* (2nd ed., 2000). He is the author of *Early Intervention Systems for Law Enforcement Agencies: A Planning and Management Guide* (2003), published by the COPS Office of the U.S. Department of Justice.

He currently serves as Coordinator of the Police Professionalism Institute (PPI) at the University of Nebraska at Omaha. The PPI is engaged in a number of projects related to police relations with the Hispanic–Latino community; early intervention systems; national standards for police auditor systems; and a comparative analysis of police accountability in the United States, Latin America, and Europe. PPI reports are available at www.policeaccountability.org.

Professor Walker has served as a consultant to the Civil Rights Division of the U.S. Department of Justice and to local governments and community groups in a number of cities across the country on police accountability issues.